AF661638

Production Technology of Minor Vegetable Crops

Related Books

Titles	Authors
— A Handbook of Laboratory Solutions	*M.H. Gabb*
— Advanced Technology in Exploration and Exploitation of Minerals *2nd Edition*	*MEAI*
— Agriculture Chemical Users Manual	*Geoff Cowles*
— Antenna and Wave Propagation	*Ashish Mathur*
— Bamboo Flowering and Rodent Outbreaks	*S. Sridhara*
— Basic Statistics *2nd Edition*	*P.K. Mohanty*
— CSIR NET: Part A (Hindi)	*Christy Varghese*
— CSIR-NET General Aptitude: A New Outlook	*Christy Varghese*
— Design of Concrete Structures	*Ramachandra*
— Design of Small Dam (Forthcoming)	*USDI*
— Design of Steel Structures (Vol. 1)	*Ramachandra*
— Design of Steel Structures (Vol. 2)	*Ramachandra*
— Drainage Engineering	*J.N. Luthin*
— Drainage Engineering: Principles and Practices	*S.K. Gupta*
— Dryland Technology *2nd Edition*	*M.L. Jat*
— Effects of Heavy Haul Trains on Kottavalasa-Kirandul Railway Line	*C.S. Sundaram*
— Engineering Mechanics Lab Manual	*A.K. Gupta*
— Engineering Thermodynamics	*R.K. Purohit*
— Entrepreneurship Development Hand Book	*O.P. Harkut*
— Environmental Engineering and Safety	*Sangeeta Raut*
— Fluid Mechanics	*Kamlesh Purohit*
— Foundation of Mechanical Engineering *4th Ed.*	*R.K. Purohit*
— Geo-Resources	*K.L. Shrivastava*
— Ground Water Manual a Water Resources Technical Publication	*USDI*
— Guidelines for Drinking-Water Quality *4th Ed.*	*WHO*
— Hydraulics and Hydraulic Machines	*Kamlesh Purohit*
— Industrial Management and Entrepreneurship	*V.K. Sharma*
— Material Science and Processes	*R.K. Purohit*
— Mechanical Engineering	*R.K. Purohit*
— Pani Bijali Bachat Jan Jagriti	*O.P. Maheshwari*
— Prabandh Evam Ughamita (Management and Entrepreneurship) *2nd Ed*	*V.K. Sharma*
— Revamping Architectural Education	*Rajesh Sharma*
— Space Technology and GIS for Disaster Monitoring and Mitigation	*T.S. Chouhan*
— Steel, Timber and Masonry Structures	*Ramachandra*
— Water Quality Standards Hand Book *2nd Ed.*	*USEPA*
— Water Treatment Manual Disinfection	*EPA*

Production Technology of Minor Vegetable Crops

Mukesh Topwal
Deputy Project Director
Agricultural Technology
Management Agency,
Department of Agriculture,
Uttarakhand Government,
Uttarkashi, Uttarakhand

Shruti Agrawal
M.Sc. (Botany),
D.Phil (Microbiology)
Assistant Professor at Govt. PG. College,
New Tehri, Tehri-Garhwal, Uttarakhand.

SCIENTIFIC
PUBLISHERS

Published by:

SCIENTIFIC PUBLISHERS (INDIA)
5A, New Pali Road, P.O. Box 91
Jodhpur 342 001, India
E-mail: info@scientificpub.com
Website: www.scientificpub.com

© 2022, Authors

All rights reserved. No part of this publication or the information contained herein may be reproduced, adapted, abridged, translated, stored in a retrieval system, computer system, photographic or other systems or transmitted in any form or by any means, electronic, mechanical, optical, digital, by photocopying, recording or otherwise, without written prior permission from the publisher. Any breach will attract legal action and prosecution without further notice.

Disclaimer: While every effort has been made to avoid errors and omissions, this publication is being sold and marketed on the understanding and presumption that neither the editors (or authors) nor the publishers nor the printers would be liable in any manner whatsoever, to any person either for an error or for an omission in this publication, or for any action to be taken on the basis of this work. Any inadvertent discrepancy noted may be brought to the attention of the publisher, for rectifying it in future editions, if published.

This book contains information obtained from authentic and highly regarded sources. Reasonable efforts have been made to publish reliable data and information, but the editors and publisher cannot assume responsibility for the validity of all materials or the consequences of their use. The editors and publisher have attempted to trace and acknowledge the copyright holders of all material reproduced in this publication and apologize to copyright holders if permission and acknowledgement to publish in this form have not been obtained. If any copyright material has not been acknowledged please write and let us know so that we may rectify it.

Trademark Notice: Publications or corporate names may be trademarks, and are used only for identification and explanation in bonafide intent without intent to infringe.

ISBN: 978-93-88812-96-2
eISBN: 978-93-88812-98-6

Printed in India

PREFACE

India is a second largest vegetable producing country in the world. Presently Indian farmers mainly focus in major vegetables production. But the importance of minor vegetable in terms of health benefits, genetic resources, crop improvement as well to increase country rank in vegetable production can't be negligible. So that minor vegetable can play important role in increasing economy of the country. Minor vegetable crop species have a potential that needs to be exploited to advantage. Most of these crop species do not require high inputs, can be grown in marginal and degraded lands, and at same time contribute to increased agricultural production, crop diversification and a better environment. In addition, such efforts will be help in conserving and using the genetic resources of underutilized vegetable species. Future strategies of the underutilized vegetable crops and their importance among the growers and consumers developing improved varieties, having wider adaptability and developing suitable cultural practices for different crops ensuring timely input supply, creating indigenous market.

This book covers all the aspects related to cultivation of "PRODUCTION TECHNOLOGY OF MINOR VEGETABLE" crops along with questions and answers embedded in text. Before the compilation, the author has consulted number of textbooks, periodicals and journals from which matters have been heavily drawn in this book and author heartily acknowledges his indebtedness for the same. This book consist production technology of twenty six minor vegetable crops. This book is mainly intended for the horticultural courses of graduate and post graduate students of horticulture. This book encloses all the information on cultivation of minor vegetables and is presented in a systematic manner to understand the fundamentals aspects clearly and easily.

Mukesh Topwal
Shruti Agrawal

CONTENT

CHAPTER 1

Introduction and Classification of Vegetables

Introduction and Classification of Vegetables

Vegetable growing is an important branch of horticulture. But it is no longer confined to growing in a small garden as vegetables are now cultivated on a large scale commercially. The study of vegetable culture is called as 'Olericulture'. The term vegetable is applicable to the herbaceous plants or parts thereof, which are used for culinary purposes. The term vegetables includes all foods of vegetable origin, but it excludes cereal and dried seeds of pulses, however, it includes grain on the cob, potatoes and other tubers.

Vegetables enrich and diversify the human diet. They are the primary source of mineral nutrients, vitamins, secondary plant metabolites, and other compounds that support human health and nutrition. Vegetables, especially roots and tubers, can also possess significance caloric value, serving as staple crops in many parts of the world, particularly in the tropics. Although vegetables account for less than 1% of the world's plants, the genetic, anatomical, and morphological diversity of vegetables as a group is astounding (Graham *et al.* 2006; Maynard and Hochmuth 2007). Hundreds of vegetable taxa are grown for food in subsistence and commercial agricultural systems worldwide. This chapter reviews and explains the biology and classification of vegetables.

Classification of Vegetable

There are about 10,000 plant species that are used as vegetables the world over. Of these, about fifty plant species or so are commercially important. It is therefore, important to classify vegetables into groups or classes to better understand or discuss them. Vegetables that share some characteristics are grouped together. There are different ways of classifying vegetable crops but none of these methods would satisfy completely the different groups like farmers, consumers, academics etc. Broadly there are five methods of classifying vegetables. These include;

1. Botanical classification
2. Classification based on hardiness or temperature
3. Classification based on plant part used
4. Classification based on culture
5. Classification based on life cycle

Botanical classification

Botanical classification is based on taxonomical relationship among different vegetables. Plant kingdom is divided into four viz. Thallophyta, Bryophyta, Pteriodophyta and Spermatophyte. All vegetables belong to division Angiospermae of Spermatophyta. It is further divided into two classes viz., Monocotyledoneae and dicotyledoneae.

The family wise distribution of vegetables under the classes is as follows:

Monocotyledoneae

Family - Alliaceae

Allium cepa- Onion

Allium cepa var. *Aggregatum* - Multiplier onion

Allium cepa var. *Viviparum* - Top onion

Allium porrum - Leek

Allium sativum - Garlic

Allium fistulosum - Welsh onion

Allium ascalonicum - Shallot

Allium schoenoprasum - Chive

Family - Liliaceae

Asparagus officinalis - Asparagus

Family - Araceae

Dioscorea alata - Larger yam

Dioscorea esculenta - Lesser yam

Colocasia esculenta - Taro

Family - Poaceae (Graminae)

Zea mays - Sweet corn

Dicotyledoneae

Family - Aizoaceae

Tetragonia expansa - New Zealand spinach

Family - Chenopodiaceae

Beta vulgaris- Beetroot and Palak

Beta vulgaris var. *cicla* - Swiss chard

Spinacia oleracea - Spinach

Artiplex hortensis - Chakwat / orach

Chenopodium album - Bathua

Family - Asteraceae (Compositae)

Cichorium intybus - Chicory

Cichorium endivia - Endive

Lactuca sativa - Lettuce

Cynara scolimus - Artichoke

Family - Convolvulaceae

Ipomoea batatas - Sweet potato

Family - Brassicaceae (Cruciferae)

Brassica oleracea var. *acephala* - Kale

Brassica oleracea var. *gemmifera* - Brussels sprouts

Brassica oleracea var. *capitata* - Cabbage

*Brassica olerace*a var. *botrytis* - Cauliflower

Brassica oleracea var. *italica* - Sprouting broccoli

Brassica caulorapa - Kohlrabi or knol khol

Brassica napus var. *napobrassica* - Rutabaga

Brassica campestris var. *rapa* - Turnip

Brassica juncea - Leaf mustard

Brassica chinensis, B. pekinensis - Chinese cabbage

Armoracia rusticana Horse-radish

Raphanus sativus - Radish

Family - Cucurbitaceae

Cucurbita peop - Summer squash

Cucurbita moschata - Pumpkin

Cucurbita maxima - Winter squash

Cucurbita lanatus - Water melon

Cucumis melo - Musk melon

Cucumis melo var. *momordica* - Snap melon

Cucumis melo var. *utilissimus* - Long melon

Cucumis melo var. *conomon* - Oriental picking melon

Cucumis sativus - Cucumber

Praecitrullus fistulosus - Tinda

Sechium edule - Chow-Chow

Luffa acutangula - Ridge gourd

Luffa cylindrica - Sponge gourd

Lagenaria siceraria - Bottle gourd

Trichosanthes dioica - Pointed gourd / Parwal

Trichosanthes anguina - Snake gourd

Momordica charantia - Bitter gourd

Benincasa hispida - Ash gourd

Family - Euphorbiaceae

Manihot esculenta - Tapioca

Family - Fabaceae (Leguminosae)

Pisum sativum - Peas

Phaseolus vulgaris - French bean

Phaseolus lunatus - Lima bean

Vicia faba - Broad bean

Vigna unguiculata - Cowpea

Cyamopsis tetragonoloba - Cluster bean

Vigna unguiculta var. *sesquipedalis* - Asparagus bean

Lablab purpureas - Lablab bean

Glycine max - Soybean

Psophocarpus tetragonolobus - Winged bean

Tigonella foenum graecum - Methi / fenugreek

Tigonella corniculata - Kasuri methi

Family - Malvaceae

Abelmoschus esculentus - Okra / Bhendi

Family - Solanaceae

Solanum tuberosum - Potato

Solanum melongena - Brinjal

Solanum lycopersicum - Tomato

Capsicum annuum - Chilli

Family - Apiaceae (Umbelliferae)

Daucus carota - Carrot

Petroselinum crispum - Parsley

Apium graveolens - Celery

Pastinaca sativa – Parsnip

Classification based on hardiness or temperature

This classification is based on the ability of crop plants to tolerate frost. Hardy crops are those, which can tolerate frost. These are essentially winter season/cool season/temperate crops and are adapted to mean monthly temperature of 15-18°C. These crops can further be divided into two sub-groups viz., hardy or tolerant vegetables and semi-hardy or semi-tolerant vegetables. Some of the hardy crops like asparagus and rhubarb can tolerate even freezing temperature. Non-hardy crops or the summer season crops like cucurbits thrive best under high temperature (20-27°C) and are intolerant to frost. Warm season crops can further be divided into two groups viz. tender or sensitive vegetables and very tender or very sensitive vegetables. Based upon the hardiness, vegetable crops are divided into the following groups.

➢ **Winter Season Vegetables**

- **Hardy/tolerant vegetables**—these include asparagus, broccoli, Brussels sprouts,cabbage, chive, collard, garlic, kale, knol-khol, leek, onion, parsley, pea, radish, rhubarb,spinach and turnip.
- **Semi-hardy/semi-tolerant vegetables**—these include carrot, cauliflower,Chinese cabbage, celery, globe artichoke, lettuce, leaf beet, parsnip and potato.

➢ **Summer Season Vegetables**

- **Tender/sensitive vegetables**- these include chilli, tomato, snap bean and sweetcorn.
- **Very tender/very sensitive vegetables**- these include amaranth, okra, eggplant, bell pepper, tapioca, chilli, cluster bean, Lima bean, colocasia, cowpea, cucurbits, sweet potato and yam.

Classification based on plant part used

In this system, crops are classified based on their parts used for vegetable purpose.

➢ **Leaves;** cabbage, lettuce, spinach, leaf beet, amaranth, fenugreek etc.

➢ **Stem;** knol-khol, asparagus, celery, cauliflower (hypocotyl branches) etc.

➢ **Fruits;** cucurbits, tomato, eggplant, chilli, bell pepper etc.

➢ **Pods;** snap pea, snow pea, beans, okra etc.

- **Flowers;** sprouting broccoli, globe artichoke.
- **Roots;** radish, turnip, carrot, beetroot, yam, rutabaga, parsnip etc.
- **Bulbs;** onion, garlic, leek etc.
- **Tubers;** potato, sweetpotato, cassava, taro.
- **Seeds;** pea.

Classification Based on Culture

This is the most convenient and widely used system of classification of vegetables. Vegetables having similar cultural requirements are grouped together and placed in one group. For eg., crops belonging to group Cucurbits are seed propagated, direct sown, trailing and vigorous growing, cross pollinated and the cultural practice are almost same. Based on their cultural and climatic requirements, vegetables are divided into the following groups;

- **Cole crops;** these crops belong to the family Cruciferae and are also called crucifers or Brassicas. These are winter season and transplanted crops. The crops include cauliflower, cabbage, knol-khol, Chinese cabbage, sprouting broccoli and Brussels sprouts etc.
- **Leafy vegetables;** all vegetables belonging to this group are direct seeded crops and include spinach, leaf beet, coriander, fenugreek, amaranth, Swiss chard etc.
- **Salad vegetables;** these crops are mainly eaten raw and include lettuce, celery, chicory and parsley.
- **Root vegetables;** these crops have prominent and fleshy underground structures and are direct sown winter season crops. These include radish, carrot, turnip, beetroot, parsnip, rutabaga etc.

Classification of Vegetable Crops

- **Cucurbit crops;** these crops belong to Cucurbitaceae commonly known as the gourd family. The plants have tendrils and produce fleshy fruits. These are direct seeded summer season crops and include melons, gourds, cucumber, pumpkin and summer squash.
- **Solanaceous crops;** these belong to Solanaceae commonly known as the night shade family and are summer season transplanted crops. These include tomato, chilli, bell pepper and eggplant.
- **Pea and beans (pod vegetables);** these are legume vegetables and belong to Fabaceae or the pea family. These are directly seeded vegetables and include pea, French bean, Dolichos bean, broad bean, cluster bean, Lima bean, winged bean and cowpea.

- **Bulb vegetables;** these are species of Allium and belong to the family Alliaceae. These are winter season crops and include onion, leek, garlic, etc.
- **Perennial vegetables;** these crops remain in the field for more than two years and include asparagus, artichoke, Jerusalem artichoke, chayote, ivy gourd and pointed gourd.
- **Tuber vegetables;** these include potato, sweet potato, taro, cassava/ tapioca, yams.
- Okra is represented by an independent group.

Classification Based on Life Cycle

- **Annual vegetables;** pea, beans, tomato, okra, eggplant, chilli, bell pepper, cucurbits etc.
- **Biennial vegetables;** cole crops, bulb crops, root crops etc.
- **Perennial vegetables;** asparagus, artichoke, Jerusalem artichoke and pointed gourd.

CHAPTER 2

Arrow root (*Maranta arundinaceae*)

Family: Marantaceae

Origin: South America and the Lesser Antilles

Chromosome number: 2n= 18

The genus *Maranta* includes 20 species of perennial herbs. One species *Maranta arundinaceae* is cultivated for its edible rhizomes. It is an erect herbaceous, dichotomously branched perennial with 0.6- 1.8m in height, the inflorescence are terminal bearing few white flowers which very rarely produce the seeds. The rhizomes are fleshy, cylindrical, swollen and thickened towards the apex. According to colour of rhizomes, blue and yellow types are reported.

These include other species ofMaranta, such as *Marantaramosissima, Maranta allouya, Maranta nobilis,* as well as Brazilian arrowroot (*Manihot utilissima or Manihot palmate*), Tahitian arrowroot (*Tacca oceanica*), and East Indian arrowroot (*Curcuma augustifolia*). While some of these starches may be chemically similar to true arrowroot, it is not clear if they produce the same medicinal effects. Consumers interested in trying arrowroot are advised to choose Maranta arundinacea, which is sometimes referred to as West Indian arrowroot or simply Maranta.

Origin and distribution

Arrowroot is indigenous to tropical America and has long been cultivated in the West Indies, particularly St. Vincent, which produces about 95 per cent of the world's commercial supply. Cultivation has spread to many other tropical countries, including Brazil, India, Sri Lanka, Indonesia and the Philippines. The plant grows wild in some parts of India and it is cultivated in Uttar Pradesh, Bihar, Orissa, West Bengal, Assam and Kerala.

Importance and Scope: Maranta (*Maranta arundinacea* L) can be considered as a non-conventional raw material for starch. Thus, the objective of this work was the characterization of the Maranta starch. Arrowroot (*Maranta arundinaceae*) is an erect herbaceous plant belonging to Marantaceae family. Commonly known as 'West Indian Arrowroot' it grows to about one and a half meter height and produces long, fleshy and cylindrical subterranean rhizomes, which tastes like, corn when boiled. The bulk of the material is used for the production of starch soon after harvest. Later the starch content decline and sugar content is increased. The analysis of fresh rhizomes showed that it contain moisture 63.4%, starch 27.8% , fibre 3.0%, dextrin and sugar 2.1%, crude protein 1.6%, ash 0.9% and fat 0.2%. The leaves of the plant are used as a packing material for fish and meat. It is primarily grown for its quality starch, which is valued particularly for infants and invalids. It is also used in making various bakery products, special glue and paste, as a base for face powder, as ice cream stabilizer and in carbonless paper used for computer print outs. Arrowroot starch is also used in the treatment of intestinal disorders and also employed in the preparation of barium meals and in manufacture of tablets. The fibrous material, which remains after the extraction of starch, is used as a cattle feed or manure. In India, it is grown in North Eastern States (West Bengal, Assam) and in southern India mostly in Kerala as a rainfed crop in limited areas in homesteads. The crop is a native of tropical America and has long been cultivated in the West Indies particularly in St. Vincent, which is the main producer of arrowroot starch.

Climate and Soil: Temperature range of 20-30°C are ideal for the crop with a minimum annual rainfall of 95-150 cm but sufficient soil moisture throughout the growing period is important. Optimum yield are obtained where the rainfall evenly distributed and with short dry season. This normally grows from sea level up to about 900m but does well near the sae at elevation of 60-90 Arrow roots require deep, well drained, slightly acid, loam soil for optimum yield. It is also grows in sandy soil and partial shade is beneficial.

Varieties and hybrids: Generally yellow coloured local cultivars are grown. However, cultivars having blue rhizomes give higher yield of starch yellow coloured cultivars.

Cultivation Practices: Small pieces of rhizomes (known as bits), 4-7 cm long, having 2-4 nodes each, are planted in well-manure pits.

Suckers are also used as planting material. They are separated from clump and planted at a distance of 30-45 cm in nursery during off-season. These suckers grow to new plants, which are uprooted, and foliages are detached to keep 10 cm. long shoot with intact roots. This is used as planting material. Normally 2 clumps are planted at a distance of 45 cm. About 3 tonnes of planting material is enough for a hectare of land. Approximately 3000-3500 Kg of bits is required to plant one hectare.

Raised beds of 50 x50 cm and 15-20 cm high are prepared. The bits/ suckers are planted at 30 cm distance at a depth of 5.0 – 7.5 cm and covered with soil. Planting is done just before the onset of monsoon.

Mode of Pollination: The arrowroot inflorescences are terminal bearing with a few white flowers but seed set is rare. Propagation is made through small pieces of rhizome having several buds.

Management of water, nutrients and weeds Irrigation: Sufficient water supply in the soil throughout the growing period is required for optimum yield. The crop is normally grown as a rainfed crop. However, if dry spell occurs during the initial 3-4 months, supplementary irrigation at weekly intervals may be necessary.

Nutrients: Arrowroot usually thrives well in fertile sandy loam soil. Heavy soils should be made friable by incorporating organic matter or compost. Use of 10 t ha-1 of FYM or compost is advocated for arrowroot cultivation.

Weeds: Field should be kept clean and free of weeds during the first 3-4 months after planting. During weeding, earthing up should also be done. Mulching with green or dried leaves reduces the weeds and also has favourable effect on soil moisture and improves yield.

Harvesting: Maturity is indicated by yellowing, wilting and drying up of the leaves. The crop attains maturity in 10-11 months after planting. At this stage, the plants are dug-out and the rhizomes separated from the leafy stem. Harvested rhizomes are 20- 45 cm long and 2.5 cm thick. Small rhizomes are used for generating planting material, whereas bigger sized rhizomes are mainly used for starch production through further processing.

Yield: On an average 1 -1.5 kg of rhizomes can be obtained from a single clump. The yield of rhizomes varies from 4-7 tonnes per hectare and higher yield of 12 tonnes per hectare has been recorded under favourable conditions.

Post harvest Handling: The rhizomes normally do not eat decomposition under ordinary storage environment. At some places, rhizomes are stored embedded in dry sand layers in dark. Physical damage to rhizomes during harvest, however, enhances the chances of deterioration under normal storage.

Diseases and Pests: No major diseases have been reported for this crop. The pests of this crop are similar to other aroids and yams as described under yam.

CHAPTER 3

Artichoke (*Cynara scolymus L.*)

Family: Compositae

Chromosome number: 2n= 34

The globe artichoke, *Cynara scolymus* L., is a member of the Compositae family. The name is derived from the northern Italian articiocco and articoclos, ascribing to the flower buds both the look of a pine cone and a globe in shape.

Origin

The artichoke is native to the Mediterranean region of North Africa and southern Europe. Ancient Greek and Roman writings make reference to the delicacy of this food item. The artichoke seemed to disappear after the Romans until its rediscovery in Naples and Sicily around 1400, when it became a cultivated plant and spread to France, Spain, and other European countries. In Britain and the American colonies the artichoke never reached the status it enjoyed in southern Europe. Louisiana and California were the original sites, and commercially only California became a prime production area.

Botany

A thistle like herbaceous perennial, the artichoke renews itself year after year by means of lateral shoots arising just below the soil surface. The plant initially forms from the shoot, which develops leaves that are large, woolly, deeply lobed, and prickly. The flower buds are in large globular heads consisting of several rows of thick bracts. It is the bracts of the unopened flower bud that are used for food, together with the large fleshy receptacle. The flowers are white, blue, or violet in color.

The plant grows rapidly spreading to 1.8 m or more in diameter and reaching a height of more than 1 m. Each year the plant develops from the shoots produced at the base, a few at first to as many as a dozen in older plants. Each shoots forms a cluster or large basal rosette of leaves covered with dense white hairs from which the stem grows.

Consumption and Uses

The U.S. per capita consumption of artichokes has fluctuated widely from 220 g per capita to 650 g in the 1925-1947 periods, down to 100 g per capita currently. In 1975-1978 artichokes ranked seventeenth in acreage and twenty-second in fresh production. About one-quarter of the artichoke harvest is processed, the bulk going for fresh consumption. As in ancient times, the delicate nutty flavor of the artichoke makes it a prized dish. The fresh bud is usually cooked whole and served hot or cold with melted butter, salad dressing, or a sauce. The canned or frozen small buds or "hearts" are popular as a hors d'oeuvre.

Major Cultivars

The only commercial cultivar is Green Globe. A number of other cultivars are grown in home gardens throughout California and other suitable areas such as Louisiana.

Production Region

The most adaptable production region in the United States is the cool coastal region cantered around Castroville in California, where frost is seldom a factor and the plant grows unimpeded the whole year.

Soil and Fertilizer Requirements

Although adaptable to a wide range of soils, the artichoke prefers a very deep, fertile, well-drained soil with high OM content. Up to 25 tons of manure per ha should be incorporated at the end of each cropping year. Nitrogenous fertilizers are applied if tissue tests show a decline in N just before the main harvest period begins.

Cultural Practices

Artichokes are propagated by side shoots that increase in numbers as the plant ages. These shoots are planted 1.2 m apart in the row with 1.8 m between the rows to permit adequate room for growth. The end of the growing cycle each year sees the plants cut back to the soil level. At one time the plant debris was burned; a better practice is to put the trash between the rows and rot till it back into the soil. The normal economic life span of the crop is three years until replanting, even though an artichoke plant can survive for fifteen or more years. Although the artichoke is a heavy seed producer, growers prefer to propagate the side shoots of plants with a known performance record.

The artichoke planting is irrigated three to five times per season beginning with the new planting or with the start of each production year. Because of the spacing between the rows, it is possible to clean cultivate initially. As they grow the plants shut out the sun required to initiate weed growth. Herbicide control may be necessary, however, to keep down excessive annual weed growth.

Pest Control

The local USDA or state or county agricultural advisor's office should be consulted as to current recommendations on registered chemical control measures that can be used.

Weeds: Between-row cultivation helps to eliminate weeds, and as the crop matures, the plants shut out the sun required to initiate most weed growth. Nevertheless the control of quick-growing annual weeds is essential. Semisterilant contact materials are being used for long term effective weed control.

Insects: The artichoke is particularly troubled with aphids, which also may carry viral diseases. Plume moth larvae attack the bracts destroying the plant base. Consultation with local extension specialists is necessary to ascertain which chemicals have EPA clearance and residual carryover tolerances.

Diseases: There appears to be no serious disease attacking the artichoke other than viral diseases spread by aphids.

Harvest and Storage

Artichoke harvesting occurs every month of the year. The fall harvest peaks in November and December and following the winter rains harvest begins in earnest in March, peaking in April and tapering off in May. The artichoke is harvested by hand cutting the stem below the base of the bud. Harvests are at first cut every 10 days, every four days in the peak period. After harvesting, the buds are separated into sizes usually determined by bud count per shipping package-large, 36 to the box; medium, 42-63; small, 72-96; and heart and baby size, 108 or more. At the consumer level artichokes are sold loose or put in pulp trays over wrapped with heat shrink film.

Artichokes are not stored for long periods. Since the buds are high heat-of-respiration generators, every effort must be made to hydro-cool them rapidly to reduce field heat to a holding temperature of 0°C. Since the bracts are very subject to rapid moisture loss, it is essential to maintain at all market points a high humidity condition, either by constant top icing or by refrigerated shelves with fine sprinkling capability, to keep the artichoke quality at a peak level. Store shelves should be refrigerated to hold artichokes at about 4°C for best quality.

CHAPTER 4

Asparagus (*Asparagus officinalis L.*)

Family: Liliaceae

Chromosome number: 2n= 20

Asparagus *officinalis* L. var. *altilis* is a member of the Liliaceae (lily) family. The common name and the latinized name are identical. *Officinalis* means "medicinal." According to Ehlert and Seeling 1966 the Latin *a* from Greek (to) and *sparasse* (tear), refer to the strong prickles on some species. The Greek *asparagus* means to "sprout" or "shoot." The English changed the name *sparagrass* (sparrow grass), and within the modern produce business the name is just plain grass. The tender shoots called spears are used as vegetables and in preparation of soup. It is also eaten as *salad.* The canned and frozen spears are also use used as processed foods. The tender shoots contain a white crystalline substance, asparagines, which are used in medicine as diuretic in cardiac dropsy and chronic gout. It has a good potential for export as fresh vegetable.

Origin

Although authorities have argued as to the exact place of origin of asparagus, all agree that the eastern Mediterranean region is the most likely. The asparagus was first noted for its medicinal properties and has been in cultivation for at least 2000 years. Roman writers set down instructions on how to grow asparagus, which even then was prized as a food crop. These early writings indicated a preference for the forms of asparagus grown from seed collected from the wild form. Obviously, since asparagus plants are found growing wild, birds have readily dispersed asparagus seed. Wild growing A. *officinalis* plants are to be found throughout Canada and the United States. Although there are at least 150 species of asparagus, only A. *officinales* is cultivated for food. The asparagus came to North America with the European colonials.

Botany

The asparagus plant is a long-lived monocotyledonous, herbaceous perennial, and it is dioecious. The plant is grown for its succulent fleshy shoots (spears), which appear after a prolonged winter rest period. As in other monocotyledonous species, the crown of the plant is the critical growth center. The plant consists of underground stems (rhizomes), fleshy roots, and fibrous roots. The fleshy roots serve as storage organs and the fibrous roots as absorption organs. The fibrous roots die after each year's growth; likewise the fleshy roots die after providing nourishment for the next generation of spears. Each successive generation of roots originates from the crown of the plant and so over time it moves closer and closer to the soil surface. This phenomenon has given rise to the idea that the asparagus plant as it ages lifts or pulls itself upward. As each growing season progresses, the rhizomes develop buds that generate the spears of the following spring and summer months. From the center of the asparagus plant there is a progressive lateral growth of rhizomes and roots of several centimetres per year (more rapid growth when the plant is young). A fairly old plant (10-15 years) will have a rhizome radius of 60cm or more. In addition to their lateral growth, the fleshy and fibrous roots over time penetrate the soil very deeply.

If not cut for food, the shoots that give rise to the spears each year eventually become the ferns of the asparagus plant. As the spear ages, lignin accumulates in the pericycle region to provide support for the ferns. The usual location for the initial build-up of lignin is the butt end or the white basal stem area. The young tender green tip or bud of the spear does not possess lignin.

The ferns of the asparagus plant are cladophylls, which perform as leaves but are actually modified stems. The true leaves are the scale like structures that form at the tip of the spear and down the stem. As the plant grows, the scales give rise to the cladophylls, which in turn give rise to the flowers.

Being dioecious, asparagus produces male and female florets on separate staminate and pistillate plants. As a rule the staminate (male) plants give rise to more spears and are consequently higher yielding than the pistillate (female). The pistillate spears generally are heavier individually but fewer in number than in the staminate plants. For this reason breeding efforts are concentrated on developing super males that will produce higher, more stable yields of spears. The normal ratio of staminate to pistillate plants is 1:1. In addition there are occasional hermaphroditic plants, which may someday become the type of plant most desired. The male flowers are yellowish green and conspicuous, the female flowers less conspicious. The differences in sex can also be identified by the flower bud formation. The staminate bud is blunt at the tip and more cylindrical in shape; while the pistillate buds are pyramidal or wedge shaped and somewhat pointed at the tip. Asparagus flowers are insect pollinated-pollen does not move with the wind until the plant is dead and dried out. The flowers are individual and form a one-to-nine-seeded berry (three carpels), which when mature is round and red. If pollination is perfect, nine large black seeds develop, rounded on the back and flattened on the side with longevity of at least three to six years.

Although it has been known that male or staminate plants produced more spears, it was often a difficult or impossible task in practical terms to wait for a mixed lot of asparagus plants to show their unique flowering qualities. Consequently little was done to take advantage of this characteristic. However, with the use of tissue culture and in particular, meristem culture, a unique high-yielding staminate asparagus plant can be quickly and efficiently used to develop thousands of new clones of that particular plant.

The asparagus plant should not be harvested for at least two years and in some regions not until the third year, in order to build up reserves in the crown. From the third to the sixth or seventh year after spear harvest begins, yields will increase yearly. Some say as the spear yields level it soon begins to decline. Others indicate that productive asparagus plantings have lasted over one hundred years.

Consumption and Uses

Fresh asparagus is sold loose or in tied bundles of spears. Asparagus is processed for canning (whole, bits, and pureed for soups) and freezing. Fresh steamed asparagus is most often served with butter or a cream sauce. About 40 percent of asparagus consumed is fresh, 50 percent canned, and the rest frozen. The annual consumption per capita is about 380 g.

Nutritional Values

Like most vegetables, asparagus retains its mineral and vitamin content better when frozen than when canned. Due to the lower fiber content of the green asparagus, canning of the spears causes some appearance breakdown-cuts and tips are in the category of the better processed product. Spears clearly should be purchased fresh to get the true taste of this vegetable.

Major Cultivars

Until very recently the Mary and Martha Washington cultivars, because of their rust-resistant characteristics, were the main North American types. The *Mary Washington*, introduced in 1910, is a cross by J. B. Norton, Concord, Massachusetts of Mary (a giant female seedling selected in a bed of Reading Giant from England) and Washington (a selection from New American of unknown origin). This improved variety provided rust resistance, good spear and growth habits, and purplish-tinged spearheads with tight, pointed buds. The branching of spears does not begin until well after market length is obtainable. The *Martha Washington* is a cross of Martha, also a female selection out of Reading Giant, and Washington. Never as popular as Mary Washington, it is rust resistant.

Newer cultivars such as UC157 and UC72 are replacing the Washington types and introducing tolerance to *Fusarium oxysporum*. UC157 does not have purple pigmentation in the spears as does UC72. From Germany have come a series of Lucullus cultivars, and from New Jersey potentially a whole new group of super males. Currently, Jersey Centennial and Jersey Giant look very promising. In Canada, a series of Mary Washington progenies identified as Viking (V35), Viking Select, and Viking KB3 have been important cultivars showing improvement over older types. A major asparagus improvement program is under way in Canada to develop male and super male types uniquely adapted to Canadian conditions with emphasis on *fusarium* tolerance/resistance.

In Europe there are new cultivars (Limbros 10 and 22) adapted to culture under glass or plastic for forcing purposes. There are also new all-male lines such as Franklin F. (Limbros 126) for either white or green spear production.

Climatic Requirements

As a perennial and herbaceous species, asparagus has specific requirements that must be taken into account. Judging by the capacity of the species to establish itself, asparagus has a wide geographic compatibility spectrum in very wide-ranging environments including the seashore, the desert, southern and northern latitudes (20° to 55°N), and high elevations. These diverse environmental conditions indicate wide adaptability. The best annual re-growth of asparagus is predicted on a mean temperature range of 16° to 24°C for spear and fern production. Where fall killing frosts occur, the asparagus plant goes into a resting period, which is essential for re-growth and ongoing spear production. Where the temperature does not create frost conditions, asparagus should be dried down by withholding irrigation to permit an induced rest period equivalent to a killing frost. If this rest period does not occur, as may happen in the lower elevations especially in the southeast, asparagus production quite quickly diminishes and becomes uneconomic. In the tropics asparagus plantings are good for one or at best two annual spear harvests because of the difficulty of inducing the needed rest period. During the active growing season asparagus responds to rainfall or to irrigation water.

Production Regions

The largest utilization of asparagus is for processing. The major producing states are California, Michigan, and Washington; New Jersey is also a producer, but hectarage planted to asparagus is declining. In descending order, Washington, Michigan, and California lead in production for processing and California, Washington, and Michigan in fresh market volume. A number of northern states grow relatively small amounts of asparagus. In Canada the key asparagus processing province is British Columbia, dependent to a large degree upon imported Washington state spears. The largest fresh and total hectarage of asparagus in Canada is in Ontario, where there is a resurgence of interest in planting new hectarages of asparagus and expanding plantings to Quebec and British Columbia.

Soil Requirements

Asparagus is a long-term crop requiring careful soil preparation that will insure a long, vigorous, productive cycle of growth. With annual crops, corrections can be incorporated in a subsequent crop year. Once asparagus is planted it becomes deep rooted, especially with an open soil, and it may be in place from fifteen to twenty years. The first soil requirement for asparagus is that it be friable, open, and deep, with at least 2 percent OM. The soil characteristics must hold up for a long period of production. Although asparagus has been considered to be tolerant of both alkaline and acidic conditions, the main requirement is that the pH level be such that the movement of nutrients is not inhibited. The *P/K* balance must be maintained-this can only be done with a soil containing adequate OM content and capable of holding moisture yet not preventing good drainage. Careful soil analysis is necessary. In preparing the planting bed, all factors must be brought to a level that will provide good maintenance for asparagus growth. Ideally, prior preparation of the planting bed should include the incorporation of organic matter either by ploughing under long-standing alfalfa or other legume crops or incorporating all growth in the year immediately preceding the planting year. If green manuring is impractical, a heavy application of well-rotted manure should be considered prior to planting. The importance of careful advance planning and preparation for asparagus cannot be overstressed.

Fertilizer Requirements

Fertilizer application prior to planting and subsequently every two years or so should be determined by soil test analysis or tissue analysis or both. The particular elements required are P and K. It is useful to have an initial high application of P and K placed at the bottom of the planting furrow in a band. Unlike N, these two elements if banded will be available for a long time. On the other hand annual application of N is an essential practice, one-half before the spear harvest begins and the other half following harvest. When it becomes necessary to apply P and K during the life of the asparagus planting, these elements should be applied by placing a band of fertilizer over the planted row rather than by the broadcast method. The problem with aging of asparagus plantings is that it becomes increasingly difficult to pursue a standard fertilizer application pattern. If the soil pH falls to 6.0 or lower, the asparagus planting must be given a lime side dressing application to counteract the acidifying effects, which tie up P and K.

Minor Elements: For the most part asparagus does not have any minor element problems. Under some conditions iron-induced chlorosis may occur; this problem can readily be corrected with an appropriate iron sulphate spray.

Cultural Practices

Depending on the climatic circumstances asparagus plantings can be established either in the spring or fall, using several methods.

The traditional method of establishing asparagus was to raise nursery rows of one- and two-year-old plants to be pulled and replanted later in permanent rows. The transplanting operation at least provided insurance against slow germination, excessive weed competition, and uneven stands. An even earlier method was to dig older crowns, divide them, and replant the crown pieces. It is well established that this latter technique was not conducive to high spear production over a number of years.

There was a sense at one time that the cheapest and most direct way to establish a planting was to use the direct (precision) seeding approach. One of the difficulties was in arriving at a workable combination of deep furrowing and a seeder that would establish the plants deep enough to assure the healthy development and continued vigor of the crown. The other factor was the difficulty during germination and emergence created by weed competition. In the late 1960s the application of clonally propagated plants derived from tissue-culturing made it practical to go back to transplanting. Highly uniform asparagus plants could be placed in deep furrows with appropriate spacing in and between rows to encourage maximum stands and produce high yields. As the young asparagus plant grows and becomes established, the deep planting furrow is gradually filled in. While the roots are still confined, an eventual slight hilling effect over the row will insure a good cover over the crown during the life of the planting.

Whatever the method used, it takes either three years from seed or two years from transplanting before the first spears can be harvested. Under certain favorable circumstances, especially with midsummer seeding, it is possible to begin the first harvest in two years and some months. With judicious cutting, asparagus continues to gain strength and vigour up to the sixth or seventh cutting season and then levels off, depending upon the strain grown. It is quite possible for asparagus with first rate soil and fertility levels to remain productive for fifteen or more years, especially if the planting can be kept free from diseases.

If the asparagus planting is kept free from water stress, particularly during the harvest periods, subsequent drought stresses do not materially influence yields of asparagus spears the following year. The performance of asparagus will be affected, however, if the planting bed is too shallow or low in nutrients or if the spear-cutting season is too long. The more northerly asparagus is grown, the shorter must be the cutting season. In California spear cutting from mature stands in some areas may be extended up to four to five months; while in Michigan and Ontario the season is four to six weeks, at first and six to seven weeks thereafter. In western Canada harvests are reduced by a week or two to permit greater fern growth to restore rhizomes for the next harvest. Strong fern growth activity insures a good rate of photosynthetic activity generating carbohydrate storage in the rhizomes. Additional N applied with supplemental watering, together with good growing temperatures, will compensate to some extent for heavy spear harvests.

Irrigation and Drainage

During the establishment and maintenance of the asparagus planting, the roots should not be deprived of moisture for very long. In the key asparagus production regions, irrigation is usually provided sometime during the cutting period. In the dry western regions irrigation is a constant source of moisture. It is usually applied by gravity between beds in a sub irrigated manner or by furrows rather than by applying water by fixed overhead sprinklers or pivot sprinklers. As the season advances, especially in the more humid eastern regions, too much sprinkler irrigation sets the stage for foliar disease attack. Most regions will need supplemental irrigation between rainfalls, and others will need irrigation on an ongoing basis to provide enough moisture for asparagus. If rainfall is lacking, it is usually a good practice to irrigate in the fall before the plantings go into the winter rest period. Although a good moisture reserve is required to prevent moisture stress, it need not be supplied by irrigation alone. If adequate incorporation of OM has taken place, most soil types will be able to maintain the asparagus planting without water stress due to the deep penetrating feeder roots.

Asparagus roots must be kept free of too much moisture. Most western soils permit good water percolation, which removes excess water. In the humid east, tile drains will prevent unwanted water build up. It is of great importance to plant asparagus on level land, or if the land is rolling keep the asparagus away from the low areas.

Pest Control

As a perennial crop, asparagus requires systematic protection from the inroads of weeds, insects, and disease. As with all crops, regional recommendations are developed that will guide the asparagus grower in controlling pests.

Weeds: Besides competing for water, light, and nutrients, weeds in asparagus harbour undesirable insect pests. The control of weeds is therefore a major production effort. The obvious non-chemical approach is to begin with clean land followed with strict vigilance in mowing to keep weeds from going to seed along the asparagus field perimeters. Similarly, weeds must be carefully managed during the spear-cutting period. A very detailed account of asparagus weed control has been prepared by the Cornell Agricultural Experiment Station staff (Cornell1987) and the Canadian government (Ontario Ministry 1987). In particular annual and perennial grasses, Canada thistle, and in some places pigweed and Russian thistle are serious problems.

Insects: Particularly serious asparagus insect problems include asparagus beetle, twelve-spotted asparagus beetle, aphids, pigweed case bearer, cutworm, tarnish plant bug, thrips, and in some areas garden symphylans. Local recommendations for chemical control of insects must be sought before applying chemicals in particular regions.

Diseases: Without resistance or control measures, major diseases such as asparagus rust would have brought about the disappearance of the cultivated forms of asparagus long ago. In Canada *fusarium* wilt is a very serious problem, especially in the humid parts of Ontario and eastern Canada. The third major disease is *Botrytis cinerea*, usually associated with insect damage to the ferns. If soil fumigants and/or sprays are to be used to control asparagus diseases, local recommendations must be strictly adhered to. Following harvest, asparagus spears are very subject to market diseases such as soft rot *(Erwinia cartovora)*, fusarium stem rots, and phytophthora rot. The latter two diseases can be effectively controlled by holding the asparagus at 0° to 2°C during this period.

Harvest and Storage

Depending upon how an asparagus planting gets started-by direct seeding or transplanting, the earliest cutting should begin around the third season following seeding or planting. Cutting may be adversely affected by sudden spring hot spells that induce rapid spear production and prematurely shorten the harvest season. Harvesting for the fresh market is still a hand task. Spears are cut with a knife or snapped off by hand when they are about 13 to 18 cm long including the white butt ends. Cutting is repeated over a five to six week period-longer in California-as frequently as the spears reach the desired length. Over cutting the first year may do long-term damage to the potential life of an asparagus planting. When asparagus is grown in the tropics, the spear cutting begins sooner and will only last for one or at the most two seasons, because the plant is unable to store enough carbohydrates to sustain rhizome development beyond a very short period.

In Michigan, methods of mechanical or mechanically aided harvest have been explored. One method has been to provide harvest aids with the cutters sitting just above the row, cutting the spears at the appropriate height and returning as often as an effective harvest can be accomplished. Another method mounts a cutting bar on the front-end loader of a tractor and slides the cutter and catching box at ground level over the asparagus spears at a fairly fast rate. Unfortunately debris and rubbish also get gathered up. The harvest produces random lengths of spears that must be quickly taken to a receiving and cleaning area to be sorted and selected for various types of packaging. This latter method is quick but lowers potential yields of marketable spears and reduced yields need to be balanced against a much lower harvest labor cost.

If asparagus has been over cut in any given year the result is usually smaller and fewer spears the following year. To correct this situation for at least one season, the cutting season should be shortened to permit a longer, fuller fern production season to replenish the crown shoots.

Some asparagus marketing disorders will be briefly touched upon.

- ***Chilling injury.*** Damage occurs if the spears are held too long in the 0°-3°C range, especially the lower level. The spear tips become dull and greyish looking, and limp. Asparagus spears should not be held longer than two weeks, but if they are, the best holding temperature is 2.2°C. If held at 4.4°C decay begins to take over.
- ***Bending and elongation.*** While in storage, if spears lie horizontally for more than a day, they bend upward. Elongation occurs if spears are left standing in free water at temperatures above 4.4°C; the amount of elongation increases as temperatures rise.
- ***Feathering.*** Asparagus bracts "feather" or separate as a normal consequence of growth. It is also a sign of senescence. This occurs in the field if spears are left uncut too long and in storage when temperatures are too high and RH is low.
- ***Low O_2 injury.*** Asparagus can be damaged by lack of oxygen only if pre-packaged with improperly ventilated plastic film. Usually this is a sign of high respiration, which could have been checked by holding the spears at cool temperatures.

CHAPTER 5

Bread fruit (*Artocarpus altilis*)

Family: Moraceae

Chromosome Number:

Origin

Breadfruit (*Artocarpus altilis*) is originated from *A. camansi* Blanco and *A. Mariannensis* Trecul. *A. camansi* Blanco (bread nut) is native to New Guinea, Moluccas (Indonesia) and the Philippines. Around 3000 years ago, breadfruit was first cultivated in the western Pacific and spread throughout the tropics by migrating Polynesians where it has been cultivated widely by the Pacific islanders. Several seedless Polynesian breadfruit varieties and bread nut from New Guinea were introduced in the Caribbean in the late 1700s, where the breadfruit is considered as food for the poor people. Later on, it has been distributed widely in Central and South America, Africa especially in Senegal, Ghana and Liberia, India mainly in the coastal regions of Karnataka and Kerala, Southeast Asia, Malaysia, Madagascar, Maldives, Seychelles, Indonesia, Sri Lanka, Northern Australia and South Florida (Deivanai and Bhore, 2010).

Nutritional value

The most economically important product of the breadfruit tree is the fruit which is a nutritious, high energy food source. A 100 g edible portion of boiled breadfruit flesh provides 114 kcal, mainly from carbohydrates, 4.9 g of dietary fibre, and appreciable amounts of minerals such as calcium, the B vitamins and essential fatty acids. Breadnut seeds are also highly nutritious because they have a high content of both carbohydrate and protein. A 100 g edible portion of dried seed contains 76.2 g carbohydrates and approximately 8 g fat, 17 g protein, 3.2 g dietary fibre and useful amounts of minerals, including potassium and phosphorus.

Soil conditions

Breadfruit and breadnut trees can grow on a wide range of soil types but they prefer sufficiently deep, well-drained and moisture retentive soils. Soils that are prone to water logging or seriously eroded and shallow should be avoided. The soils should have fairly high organic matter content, adequate nutrient status and a pH of 6 to 6.5.

Climate

Breadfruit and breadnut are tropical crops that grow best at temperatures between 210 and 320 C; growth is retarded at lower or higher temperatures. Rainfall both crops thrive best in locations where the annual rainfall receipt is 1500 mm to over 3000 mm and well-distributed throughout the year. Breadfruit has a higher water requirement because of its shallower root system and larger canopy. Breadnut can tolerate drier conditions better although breadfruit will grow where rainfall receipt is as low as 1000 mm annually but well distributed. This is also an important requirement for good growth. Light shade is useful to prevent wilting, stunting and possible death of newly planted trees because the large leaves lose moisture rapidly but the root system is not sufficiently developed to absorb adequate quantities of water. The trees compete strongly for light, by growing quickly to produce a tall, trunk with high branches. Therefore, the trees should be exposed to full sunlight to encourage a lower canopy to develop. Wind Sheltered locations are preferable because persistent wind can cause wilting, leaf damage, a lopsided canopy due to more growth on the side that is away from the wind, and stunted growth. Hurricane-strength winds strip off leaves, break branches and may uproot breadfruit trees.

Land Preparation

Prepare the land till the fine tilth stage by giving couple of ploughing. Land must be free from weeds and other crop residues. For increasing soil fertility, well rotten FYM mixed during ploughing.

Propagation

Bread fruit is propagated by root cutting. Root cuttings of 2.5 cm in diameter and 20 cm in length planted horizontally.

Cultivation

Planting of bread fruit is done during December. Pits of 60X60X60 cm should be dug at the distance of 10 to 12 m distance. Well decompose organic manure is apply @ 25 kg per tree and N:P:K in the ratio of 7:10:5 at 1-2 kg/plant.

Irrigation

Irrigation should be done just after planting. Later irrigation is depends on soil moisture and rain fall condition. During hot months or dry weather, a subsequent irrigation is done.

Harvesting

Bread fruit start bearing of fruit after 3 to 6 years of planting, depending upon soil condition, planting material and weather conditions.

Yield

A fully grown tree may yield about 600 to 2000 fruits per tree. An average fruit weight is about 1 to 5 Kg.

CHAPTER 6

Broad Bean *(Vicia Faba L.)*

Family: Leguminosae

Chromosome Number: 2n= 12, 14

Origin:

Vicia Faba, one of the oldest domesticated food legumes, has been cultivated for at least 5000 years and has since spread throughout Europe, North Africa, and Central Asia. It spread to China over 2,000 years ago via traders along the Silk Road, to South America in the Columbian period and more recently to Canada and Australia. The exact geographical origin of *V. faba* is still debated, although the high rainfall regions of Central Asia and Mediterranean region have proposed as possible origin. No wild progenitor has been found and major differences exist between V. faba and other species belonging to the Narbonensis complex (*V. narbonensis, V. galilea, V. johannis* and *V. hyaeniscyamus* (Zohary and Hopf, 1973). Raina and Ogihara (1995) reported chromosome number 2n = 12 for *Vicia faba* and 2n = 14 for species of the Narbonensis complex. All attempts at inter-specific hybridization between *Vicia faba* and these other species have been failed (Ramsay and Pickersgill, 1986).

A subspecies, *V. faba* paucijuga, presently found from Afghanistan to India, is a primitive form. The main characteristics of this ancestor are short stem, small leaves, very small seed (1000 seed weight lower than 250 g) and preferential autogamy. Large seeded types (*Vicia faba* major) with 1000 seed weight greater than 1 kg have developed in South Mediterranean countries and China. These types expanded in the 16th century toward Mexico and South America. Small seeded types with 1000 seed weight less than 500 g (*Vicia faba* minor) are found in Ethiopian area and have been favoured by North European agriculture. Medium seeded types (*Vicia faba* equina) have developed throughout Middle-east and North Africa with major concentration in Egypt. Therefore, Cubero (1974) argued for considering four equivalent subspecies: paucijuga, minor, equina and major.

Botany

Faba bean is an annual herbaceous plant with coarse hollow stems that can reach heights of 2 m. It has large pinnate leaves, consisting of two to six leaflets. Winter cultivars have usually higher branching (4-6/ plant) than spring ones (1-2/ plant). The plant also has a thick tap root up to 1 m long with numerous lateral/ secondary roots and bears nodules containing the nitrogen-fixing bacteria Rhizobium leguminosarum bv viciae. Flowering in early varieties begins from about the 5-7th leaf bearing stem node and up to the 15th or higher node in late varieties. Flowers are born in clusters (inflorescences) comprising 3-8 flowers (varying with variety) in the angle between leaf and stem (axil) at each node. Inflorescences form in succession up the stem as each new node is produced, over a period of 6–10 weeks, or about 15 flowering nodes. One to four pods usually develop from each flower cluster. Flowers, 2-3 cm long at anthesis, have a typically papilionaceous structure. They can be completely white, brown or violet. In most cases they concentrate their colour on black or brown melanin spots on the wings. Flowering finishes when daytime temperatures approach 30 °C, after which an extra few leaf bearing nodes are produced. Pods are short and erected in *V. minor* and *V. paucijuga* types (3-4 ovules/ pod), and long and hanging in *Vicia* major types (8-12 ovules/ pod). *V. equina* types are intermediate having 4 to 8 ovules/ pod. Like many legumes, only 15 % flowers develop pods. The seed-set is frequently higher on the bottom podding nodes of a stem and in proximal pods of a raceme. Within the pod, abortion is reported to be greater at position proximal to the peduncle. As pods mature they turn black, as eventually do the stems and leaves of the plant.

Soils

The ideal soil pH for growing faba bean ranges from 6.0 to 6.5. Fava bean does best on welldrained silt-loam soils. Lime should be applied if pH level is below 5.6. Sandy loams are also suitable but require more frequent irrigation. The crop may suffer moisture stress in soils that dry quickly because faba bean has relatively shallow roots. Faba bean seems to be tolerant to short period of water-logging.

Climate

Faba bean is a cool season legume therefore it germinates and grows well under cool soil conditions. Fields may be planted as soon as they can be worked on. The optimal temperature for plant growth is 59-68°F, especially during the stages of flower and pod development. Faba bean flowers will abort if temperatures exceed 80°F and they are particularly sensitive to hot, dry conditions during pod development. Thus, prolonged cool weather in the spring is ideal for development of pods. It is a hardy plant and withstands low temperature as low as 4^0 C.

Cultivation

Broad bean requires low temperature for its cultivation. Crop is tolerant to water stress due to protein accumulation. It tolerates water-logging and salinity. It is shown during July to August and November to December. Seeds @ 25/ha are sown in shallow channels spaced at 75 cm. In each channel, seeds are sown in double rows at spacing of 25 cm. Cultivation practices are similar to that of pole type pea. Plants are stacked with wooden sticks.

Application of fertilizers

Apply 25 of FYM and 50 kg P and 25 kg K/ha as basal dose. 25 kg N and 25 kg of K/ha are applied between 20-25 days after sowing and application of remaining 25 kg of N is done between 40-45 days.

Irrigation

Fava bean requires an average annual rainfall of 16 inches during its growth. Otherwise, it needs supplement for irrigation. Soil type does not affect the amount of total water needed but does dictate frequency of water application. Lighter soils need more frequent irrigation, but less water applied per application. The two critical irrigation times are flowering and pod set stages, both of which are very sensitive to drought stress conditions. If using overhead irrigation, as opposed to drip irrigation, it is advisable to irrigate when it's sunny to allow Figure 3. Honey bee is pollinating fava bean flower 10 plant leaves to dry off quickly before the dun goes down. This will lessen the potential infection of diseases.

Harvesting and Yield:

Tender pods are harvested at green shell stage, 3-4 months after sowing. Under favourable conditions, crop yield is 7-10 t. of green pods and 1-1.2 t. of green beans/ha.

CHAPTER 7

Ceylon Spinach, Indian spinach, Malabar nightshade (*Talinum triangular* (Jacq) Wild)

Family: Portulacaceae

Chromosome number: 2n= 48

It is soft leafy vegetable grown in tropics. Being a shad loving species, it is suitable for cultivation under other crop canopy. The leaves and tender shoot are used as vegetable.

Origin and distribution

The crop is native to Barzil and later it was introduced to Java from Brazil. At present it is being cultivated as minor vegetable in India, Malaysia, Indonasia, Arabian countries, USA and West Indies. The genus *Talinum* which has nearly 50 other species of annual or perennial herbs are more or less succulent, sometimes woody at base. The rslated species in which leaves are eaten as vegetable are *T. arnotti, T. caffuram. T. patens, T. portulecefolium, T. crassifolium, T. cuneifolium and T. indicum.* Today, it is grown widely in the tropics as a perennial and in warmer temperate regions as an annual. Malabar Spinach plants are green *(Basella alba)* or purplish *(Basella rubra)* vines with thick fleshy leaves exhibiting one of two leaf forms: closely spaced oval to rounded leaves; or, more widely spaced heart-shaped leaves

Botany:

The plant is a herbaceous perennial, erect or decumbent and 35-90 cm in height. The stems are green or purplish. The leaves are somewhat fleshy, ovate or heart shaped 5-12 centimeters (cm) in length, stalked; tapering to a pointed tip and heart shape at the base. The spikes are axillary, solitary and 5-29 cm in length. The flowers are large, pink and about 4 millimetres (mm) long with recemes inflorescence. The fruit is fleshy, stalkless, ovoid or nearly spherical, 5-6 mm in length and purple when mature. Seeds are black, small, smooth and mature rapidly after flowering. Short-day condition favour flowering and fruiting. The crop is highly self- pollinated.

Nutritive value:

Shoots of alugbati per 100 grams (g) edible portion contain water (91 g), protein (2.1 g), fat (0.3 g) carbohydrates (3.9 g) and fibre (1.3 g). The energy value is approximately 112 kJ/100 g. The protein content is relatively low compared to other greens. The vitamin and mineral contents vary widely as follows: vitamin A, 1686-6390 IU; vitamin C, 29-166 mg; calcium (Ca), 16-117 mg; and iron (Fe), 1.2-3.1 milligrams per 100 gram edible portion.

Soil and climate:

The crop prefers well drained soil rich in organic matter. It is grows up to an altitude of 1000 m. Most humid condition favour optimum growth and development of the crop.

Propagation:

It is propagated through seeds or herbaceous stem cutting.

Cultural practices

Three to four cuttings or seeds are planted per hill at a distance of 20 to 30 cm between hills in furrows spaced at 50 to 75 cm apart. If cuttings are abundant, 7 to 10 cuttings are planted in a hill with distance 40 to 50 cm between hills. It grows well even without fertilizer but yield will improve with a liberal amount of compost. Irrigation is done by watering or flooding the field.

Seedlings raised in container are transplanted in raised beds when 5.8 cm tall. They are planted at a distance of 30 cm apart. As a basal dose, farmyard manure is applied at the rate of 5kg/ 10m2. Foliar application of urea (1.5%) is also beneficial.

Harvesting:

The tender shoots are harvested 6- 8 weeks after planting. During short- day condition, flowering occurs and seeds develop very rapidly affecting the cooking quality of the vegetable. Shoots about 15- 20 cm length are cut when leaves are fully developed. Terminal shoots are removed at first harvest to encourage lateral shoot development for subsequent harvests. When the branches from a dense mass, it is better to replace the old plants with new planting.

Pest and Disease Management

Ceylon spinach is very susceptible to rock-knot nematodes, but is free of disease and pest problems due to its very thick leaf cuticle. Leaf spots caused by *Cercospora* and *Acrothecium* sometimes occur.

CHAPTER 8

Chekurmanis (*Sauropus androgynus*) Meer

Family: Euphorbiaceae

Origin: Indo Burma

Chromosome Number: 2n= 46

Plant Description:

The plant is native of India and Burma region. it is found in the Sikkim, Himslayas, Khasi and Akra hills at 1200m elevations and in the Western ghats of Kerala from Wynad northwards at an attitude of 300- 1200m. It is reported to be introduced into Kerala from Malaysia in 1953 and is known in rnalayam as 'Madura keera'. The plant is a slow growing glabrous perennial shrub and attain a height of 2-3.5m. Its main branches are terete and flaccid. Lateral branches are thin. Leaves are ovate, oblong, sessile, alternate, membranous and short stacked. The inflorescence is auxiliary with small reddish flowers. Flowers are monoceious, minute, auxiliary, pedicelled and clustered. Fruits sessile, white or pinkish-white, 0.2cm in diameter with a fleshy epycarp. The crop is highly crosspollinated and entomophillous because of photogynous and monoecious nature of flower. The genus Sauropus consists of several other species like *S. assimilis* Thw, *S. netroversus* Wight, *S. rigidus* Thw and *S. qudrangularis* Mwell.

Nutitive Value

The leaves of chekurmanis are highly nutritious, being a very rich source of micronutrients and protein. In this study, chekurmanis leaves were evaluated for protein and micronutrient content at different stages of maturity. Maturity of the leaves did not alter the rich content of protein which remained around 22.0 g/100 g. Total dietary fibre content of the leaves ranged between 34 and 36% at different stages of maturity. Fully matured leaves contained a significantly higher amount of calcium, as compared to the tender leaves. The iron content of these leaves ranged from 3.89

to 4.50 mg /100 g, while the zinc content of the same was between 1.26 and 1.48 mg/100 g. The bioaccessibility of iron and zinc was significantly higher in the tender leaves. The tender and partially mature leaves of chekurmanis contained 74 and 69 mg/100 g of niacin, respectively. The β-carotene content of the leaves ranged from 7400 to 9250 µg/100 g, while that of vitamin E ranged between 17.6 to 15.6 mg/100 g. Chekurmanis leaves contain an alkaloid, the content of which was significantly higher in mature leaves (1740 mg/100 g) as compared to tender leaves (1439 mg/100 g). The alkaloid content of chekurmanis leaves was significantly reduced upon pressure cooking. This is the first report on the nutrient content of chekurmanis leaves at different stages of maturity, as also on the bioaccessibility of minerals.

Cultivation

It comes up well in all types of soils. The growth and yield are high when they are grown in rich, well drained sandy loam or semi-laterite soils. A. warm humid climate with good rainfall is best suited and its grows luxuriantly at a lower elevations at 500m above MSL though plants are found at higher elevations up to 1,200m above MSL. When the plants are raised under shade, it produces broader leaves. There are no distinct varieties or types available in this crop.

Propagation

It is propagated by seeds and stem cuttings. Seed propagated plants come to harvest little later than cuttings. Herbaceous stem cuttings (6- 12 months old) may be collected with 5-6 nodes or 20-30cm length and planted in polybags containing manure mixture. Rooting can be harvested by dipping the cuttings in 50ppm IAA/ IBA before planting in polybag. It takes 20-25 days for rooting. Usually, it is grown as hedge or fence around vegetable plots or in kitchen gardens. For planting a hectare area, about one lakh cuttings are required. The cuttings are planted in shallow furrows at least 15 days earlier to the onset of monsoon during April-May. Cuttings are planted at a spacing of 60cm between the plants. If the crop is planted in pits (30cm3 size), 5kg of farmyard manure is added. In addition to 25g each of urea, superphosphate and muriate of potash. Application of 30g of 7:10:5 NPK mixture per plant supplemented with 1 % urea spray after each clippings enhances leaf yield considerably. When the plants are about one meter height, they are tipped to develop laterals. If left untrimmed in early age, the plant reaches the size of a small tree. It can be trained as hedge or even on bowers or on trellises.

Diseases management

The plant is devoid of any serious pests and diseases except an occasional incidence of scale insects and aphids which can be controlled by spraying Malathion at 1 mi/litre. Caution should be taken not to pluck the leaves and tender shoots immediately after spraying insecticide.

Harvesting and Yield

The first clipping of succulent leaves can be done 3- 4 months after plantings, i.e. when the plant reaches about 60-90cm height subsequently in every fortnight, another harvest is possible, if the plants are manured and irrigated. Plants are usually trimmed to 1- 1.5m height to facilitate easy harvest. The tender shoots and leaves are used for culinary purpose. The annual productivity is 30-50 tonnes/ha with a per plant yield of 1- 3kg leaves per year.

CHAPTER 9

Chow-Chow (*Sechium edule*)

Family: Cucurbitace

Chromosome Number: 2n= 24

Chemical Composition: (Per 100 g)

Component	Fruit	Shoot and Leaf	Tuberized root
Water (g)	94	90	80
Calorie (kcal)	28	60	79
Starch (g)	0.2	0.7	13.6
Protein (g)	1	4	2
Fat (g)	0.2	0.4	0.2
Carbohydrate (g)	5.6	4.7	17.8
Fiber (g)	0.7	1.2	1.0
Ashes (g)	0.5	1.2	1.0
Ca (mg)	15.5	58	7
P (mg)	17	108	34
Fe (mg)	0.4	2.5	0.8
Vitamin A (mg)	5	615	-
Thiamin (mg)	0.03	0.08	0.05
Riboflavin (mg)	0.04	0.18	0.03
Niacin (mg)	0.5	1.1	0.9
Vitamin C (mg)	15.5	16.0	19.0

(According to Saade 1996, Aung *et al.,* 1990 and Verma *et al.,* 2014)

Cultivars:

Meghalaya Local, Broad Green, Pointed Green and Oval Green are the recommended varieties for this region.

Climate and Soil:

Chow-chow thrives well in moderate climate with high humidity as prevalent in north eastern region. It does not stand in extreme dry wind during summer and frost in winter. Chow-chow requires loose, well-drained soil rich in organic matter for its growth. Chowchow is slightly tolerant to soil acidity. Field Preparation In January and February, pits of size 0.5 x 0.5 xO.5m are dug. FYM is mixed with soil and pits are filled with one third by FYM and sprouted fruit is planted directly on center of pits. Pits system of growing is adapted for this crop so that the dug soil can support the vine for long period.

Cultivars:

Meghalaya Local, Broad Green, Pointed Green and Oval Green are the recommended varieties for this region.

Climate and Soil

Chow-chow thrives well in moderate climate with high humidity as prevalent in north eastern region. It does not stand in extreme dry wind during summer and frost in winter. Chow-chow requires loose, well-drained soil rich in organic matter for its growth. Chowchow is slightly tolerant to soil acidity.

Field Preparation

In January and February, pits of size 0.5 x 0.5 x 0.5m are dug. FYM is mixed with soil and pits are filled with one third by FYM and sprouted fruit is planted directly on center of pits. Pits system of growing is adapted for this crop so that the dug soil can support the vine for long period. Seed Rate 1500 fruits/ha.

Time of Sowing: Chow- chow is shown during Rainy season

Spacing: 2m X 3m

Manure and Fertilizer

The dose of fertilizer depends upon soil, climate and system of cultivation. Well rotten FYM @15-20 t/ha is applied at the time of land preparation. NPK @ 120:80:80 kg/ha added with full dose of P and K before sowing and half dose of N at the time of vining and remaining half dose of N before flowering.

Training and Pruning Training

Pandals or trellis are erected for the vines to trail over at a height of 2 meters in homestead garden, the vines are also allowed to climb on small trees. Most trellis in northeast India is constructed by bamboo about head height to facilitate walking beneath the vines for harvesting and other operation. Bower can also be prepared at the height of 5 feet placing bamboo poles followed by criss cross wire netting. Training on bower must be started just after 30cm of vine length from the ground level. Once the crop anchor, branches spread easily because of tendrils.

Pruning: There are two fruiting season in a year and vines are pruned at end of each season leaving only a small portion of about 1.5m of stem.

Irrigation

In NEH region like Meghalaya no irrigation is required as it is grown under rainfed condition, but a constant supply of water is essential for good growth.

Harvesting and Yield

Peak season for harvest is between October-December and May-June. Chow-chow should not be allowed to become over mature, but picked prior to seed development. Well grown plants give 200- 300 fruit per year. Fruit weight is about 200- 450 g.

Post harvesting

How the fruits are taken care of when harvested is of paramount importance from a commercial perspective, particularly in view of the fact that chayotes stored at ambient temperature spoil completely after 30 days (Vargas 1991). Tests carried out to date to find the most efficient ways of protecting fruit packed for export show that the use of plastic bags greatly reduces loss of humidity although it increases the incidence of pathogens such as Mycovellosiella cucurbiticola. This, however, can be controlled through the use of antitranspirants, which also reduce chilling injury (Alvarado et al. 1989). In addition, it has been suggested that tests should be carried out to determine the optimum conditions of temperature and humidity in cold chambers. Tests at temperatures of 13-14°C and 80-90 % humidity have shown that fruit begins to germinate after 15 days, drops constantly in weight and pathogens develop (Alvarado et al. 1989; Valverde et al. 1986).

CHAPTER 10

Coleus/ Chinese potato (*Solenostemon rotundifolius (Poir)*)

Family: Labiatae

Chromosome number: 2n= 64, 84

Origin:

Coleus commonly known as Chinese potato (*Solenostemon rotundifolius* Poir.), belongs to the family Labiatae, is believed to have originated in Central or East Africa, but was early spread throughout tropical Africa and into South-East Asia, including India, Sri Lanka, Malaysia and Indonesia, where it is cultivated on a small scale.

Botany:

It produces edible tubers which can be used as a vegetable and it is an important minor tuber crop grown extensively in most of the homestead garden in Kerala and Tamil Nadu. The genus Coleus belongs to the family Labiatae contains about 200 species including ornamental and medicinal plants. It is believed to have originated in Africa. It is a small, bushy annual herb with succulent stem and aromatic leaves. It grows to a height of 30- 60 cm. Leaves are opposite, dark in colour, petiole and round- to- ovate in shape. Flowers are small and arranged in racemose cymes, the two lipped calyx and corolla were uniformly greenish- brown and violet, coloured respectively. Anther and stigma are also violet in colour. It bears a cluster of dark- brownish, hetermorphus tubes. The duration of the crops is about 5 month.

Chemical composition:

The composition of the raw tubers per 100 g edible portion is: water 75.6 g, energy 394 kJ (94 kcal), protein 1.3 g, fat 0.2 g, carbohydrate 21.9 g, fibre 1.1 g, Ca 17 mg, Fe 6.0 mg, thiamin 0.05 mg, riboflavin 0.02 mg, niacin 1.0 mg, ascorbic acid 1 mg.

Climate and soil:

The crop thrives well under tropical and subtropical conditions. The crop requires a well drained fertile soil.

Variety:

No distinct variety is identified in the intensively cultivated pockets of Kerala and Tamil Nadu. The relative absence of morphological and genetic variability may be the consequence of vegetative reproduction and since the crop is completely sterile, hybridization and further improvement is not possible. Sree Dhara is a released variety in Kerala which recorded an yield of 25- 29 tonnes per hectare.

Planting material:

Propagation is done by vine cuttings. For the production of vine cutting, tubers are first planted on flat beds. One kilograms tuber is planted in one meter squire area. The tuber starts sprouting in about one week and grows to a height of 15- 20 cm in two weeks. Nursery raised with one kilogram tuber is sufficient to plant 75- 100 meter squire. The crop can be raised both under upland and low situation where there is no water- logging. The land is ploughed or dug to a depth of 15- 20cm and brought to a fine tilth. The crop can be planted either on flat raised beds or on small ridges.

Spacing:

Optimum plant population would be 55,000 per hectare with a spacing 60 x 30 cm

Intercultural operation:

A weeding and earthing up six weeks after planting along with top dressing of fertilizers is suggested. It is important to cover a portion of stem with soil to promote better tubulisation and development. One more earthing up has to be given one month after the first earthing up.

Harvest:

The crop is ready for harvest in about five months. The tubers are ready for harvesting when the leaves begin to wither, and are normally dug by hand. Harvesting cannot be delayed as the mature tubers deteriorate rapidly if left in the soil, but they can be stored successfully in dry sand or in a cool, well-ventilated shed.

Yield: Yield range is 10- 15 tonnes per hectare.

The hausa potato is relatively free from pests and diseases though Pycnarmon cribata, Phostria piasusalis and a leaf folder, Hymenia curvalis, have been reported from India as being important. These have been controlled by spraying with pesticides such as dimethoate.

CHAPTER 11

Colocasia (*Colocasia esculenta (L.) Schott*)

Family: India, south-eastern Asia

Chromosome Number: 26

Origin: Araceae

Botanical Description:

Perennial herb to 1.5 m (4 ft) tall, with thick shoots from a large corm; slender stolons also often produced, along with offshoot corms. Leaf blades to 60 cm (24 in) long and 50 cm (20 in) wide, arrowhead shaped, with upper surface dark green and velvety; leaves peltate (stalked from back of blade); petioles large, succulent, often purplish near top. Inflorescence on a fleshy stalk shorter than leaf petioles; part of fleshy stalk enveloped by a long yellow bract (spathe). Flowers tiny, densely crowded on upper part of fleshy stalk, with female flowers below and male flowers above. Fruit a small berry, in clusters on the fleshy stalk.

Climate and Soil

Taro requires warm climate with plenty of moisture, which is available in the tropical and sub-tropical countries of the world. It can as well be grown even in the warmer temperate regions. The crop has the ability to withstand waterlogged and reduced conditions, since it can transport O2 from the leaves to the roots under such conditions. It can be grown up to 1500 m elevation with well distributed rainfall of about 1000 mm during growth period.

Taro grows well on all types of soils, the ideal type being deep well drained, friable loams, particularly alluvial. It comes up well in loamy soils with a pH range of 5.5-7.0 and a combination of warm and moist climate with a mean temperature of 21-270 C. In areas where rainfall is less, a good amount of supplementary irrigation is required for successful production.

Varieties and hybrids

Colocasia has wide variability and a large number of local cultivars are grown in different parts of India. Satamukhi, Sree Rashmi and Sree Pallavi are improved varieties. White Gauriya, Kakakachu,

Panchamukhi, NDC 1, NDC 2, NDC 3, Saharshamukhi, Kadma, Muktakeshi, Nadia Local, Ahina Local Telia and Jhankhri are promising varieties for eastern zone; Satamukhi and Saharshamukhi for western zone; and Satamukhi, Sree Rashmi, Sree Pallavi and C 16 for southern India.

Propagation

Colocasia is propagated vegetatively mostly by small cormels weighing 20-25g. Healthy, disease and injury-free uniform sized planting material should be selected and stored in a cool place at least for three months before planting. One ton planting material is enough for planting one hectare. Mini-sett technology has been developed for colocasia multiplication also. In this case, mother corms are detached from healthy plants at harvest. Such selected mother corms are first cut into cylindrical pieces and then cut horizontally into mini-setts of about 10 g weight. The mini-setts are then directly planted in the field on mounds.

Field preparation involves ploughing and turning of soil followed by planting by anyone of the methods, viz., pit, mounds, ridge and furrow depending on the soil type. Among these methods, furrow planting is superior to others due to better soil moisture retention. Depth of planting does not have any marked effect on cormel yield. However, planting seed material at a depth of 5-7 cm is the most accepted practice. Optimum spacing varies for different agroclimatic conditions. In Punjab 45x30 cm or 60x20 cm, in Kerala 60x45 cm and in Uttar Pradesh 60x20 cm have been found to be optimum.

Mode of Pollination

The aroid crops are generally characterized by lack of flowering and seed set. Continuous vegetative propagation had resulted in total lack of sexual phase in many accessions. Flowering is highly erratic. Flowering and fruit set under natural condition has been observed in few diploid accessions of colocasia. In flowering accessions also, natural seed set is prevented by different structural and functional barriers. Flowering behaviour and other cytological abnormalities are also found to affect male and female fertilities and prevent natural seed set.

Natural seed set do not occur in colocasia in diploid accessions when synchronous flowering occurs in two or more clumps. The spadices appear 3 to 4 month after planting (vegetative growth). Flowering period in each clump is 3 to 4 weeks duration. In colocasia, stigma becomes receptive 24 hours before anthesis and lasts up to 36 hours. Pollen is viable for 24 hours and pollen shedding occurs one day after anthesis. Anthesis occurs in the morning and the spadix has fragrance. Fruits flies have been found to act as pollinating agents. The berry has been noticed in the buds in flowering clumps. The berry bunches mature within a month. After flowering plant continues with vegetative growth irrespective of the development of fruit bunch.

Management of water, nutrients and weeds

Irrigation: Taro is primarily adapted to moist environments, but can very well grow under a wide range of moisture regimes. It is essential to maintain the soil moisture near field capacity which when decreases appreciably should be supplemented with irrigation. Moisture may not be a limiting factor when the crop is cultivated in a paddy culture or swampy conditions. However, under rainfed or up-land condition, irrigation should be assured. Taro when cultivated during summer requires frequent irrigation at about 10 days interval.

Nutrients: Taro is highly responsive to fertilization. A fertilizer dose of 80:60:60 Kg/ha of N, $P_2O_5$and K_2O has been found economical for most parts of India. Half dose of N and K and full P should be given at the time of planting, while the remaining N and K is to be applied in two splits the first one 7-10 days after sprouting and applied a month later.Traditionally taro is applied with plenty of organic manure, especially farm yard manure (FYM) to supplement nutritional requirement of the crop. Highest tuber yield is obtained at a manurial basal dose of 12.5 t ha-1 (FYM).

Weeding: Taro field gets heavy weed infestations during the early as well as the later phase of its growth. Therefore, at least two manual weedings, coinciding with the two split applications of fertilizers and earthing up and a third one at later stage of the crop growth are needed to produce high cormel yield. Sukers/side shoot development is a common feature in taro. This may later on add to the number of tubers obtained, but the corms on these are mostly of unmarketable kind. Tuber size, yield and quality can be increased by desuckering unhealthy suckers and retaining only healthy ones.

Mulching: Mulching after planting helps in conserving the soil moisture and also checks weed growth. Green leaf or paddy husk mulching has a significant effect on increasing yield in taro.

Harvesting and Yield

Crop will be ready for harvest in 6-8 months after planting. One month prior to harvest, all the suckers may be wrapped around the base of the mother plant and covered with soil by earthing up, for arresting further vegetative growth and sprouting of tubers. After this, irrigation has to be withheld to hasten maturity. Harvesting is done by carefully uprooting the plants and the mother corms and cormels are separated. Taro is ready for harvesting when most of the leaves begin to turn yellow. Duration of the crop varies with cultivars and methods of cultivation. In India, taro is ready for harvest in about 5 -6 months. Harvesting could be done at the convenience of the farmer as no serious deterioration has been reported if the harvest is delayed over a few weeks.

Post harvest Handling

The crop matures 120-150 days after planting. Harvesting is done by digging out the corms and cormels. The mother corms and cormels are separated before marketing. Damage to tubers should be avoided while harvesting. The damaged tubers should be separated from the marketing lot and consumed within 2-3 days. The tubers selected for marketing should be spread on ground under shade for a day and the soil adhering to the tubers should be removed before packing. Tubers should not be packed in airtight conainers. Packing in *jute* bags or baskets prevents rotting during storage.

Major diseases and pests

Leaf blight (*Phytophthora colocasiae*) appears as small dark round spots on upper surface which enlarge, by coalescing together, forming characteristic rings of yellow and brown zones. Infection in petioles lead to collapse of plants. It destroys chlorophyll and reduces photosynthetic area, causing 25 -50% yield losses. Growing resistant, tolerant varieties like Muktakeshi, Jankhri, Topi -I, and Nadia local are advocated. Early planted crop (February-March) escapes the disease.

Mosaic (*Dasheen mosaic virus*) causes interveinal yellowing and mosaic resulting in deshaping and distortion of leaves in severe cases. Use of healthy planting material and rouging infected plants is advocated.

Aphids (*Aphis gossypii*) causes yellowish speckling and curling of leaves. In severe cases sooty mould develops, reducing vitality of leaves. Major pests of taro and elephant -foot yam can be controlled by spraying insecticides such as Dimethoate, Quinalphos or Fenthion (0.05%) at 30 and 60 days after planting.

Thrips (*Caliothrips indivus*) causes characteristic silverish white specks with black dots on the leaves and drying foliage.

Leaf-eating caterpillars (*Spodoptera litura*) causes economic defoliation ranging from16-80%. Young caterpillars feed on leaves by scraping gregariously. Older caterpillars feed voraciously on the entire leaf.

Scale insect (*Aspidiella hartii*) causes shrivelling due to desapping and feeding. In case of severe infestation, the plant shows withering and drying, reducing vigour and vitality of plants. The infested tubers lose quality and viability. Use of pest-free seed tubers and dipping seed tubers before planting in 0.1% Dimethoate is advocated.

Arecanut beetle (*Araecerus fasciculatus*) bores into tubers and converts it to powder causing loss in weight and quality. Storing uninfested tubers in white sand or saw dust prevents infestation. Seed tubers can be stored in a mixture of sand and Carbaryl dust (10%) in a 100:1 ratio.

Mealybugs (*Pseudococcus citriculus; Rhizoecus sp.*) Nymphs and adults and their powdery meal cover the tuber surface. It leads to poor germination and also affects cooking quality. If pest infestation is severe,treatment spray of Dimethoate (0.05%) is effective in checking its further spread.

CHAPTER 12

Elephant foot yam (*Amorphophallus paeoniifolius*)

Family: Araceae

Chromosome Number: 26, 28

Origin: India

Importance and Scope

Elephant foot yam (*Amorphophallus paeoniifolius*) is basically an underground stem tuber and is gaining popularity due to its yield potential and culinary properties. It is rich in nutrients and is a delicacy as a food. The tubers are also used for preparing many ayurvedic preparations. Amorphophallus has an important position among tuber crops, since it happens to be highly remunerative to farmers.

Amorphophallus is indigenous to tropical Asia and Africa. It is widely distributed at low and medium altitudes in the Philippines, Malaysia, Indonesia, Sri Lanka and the South-East Asian sub-continent. It is a popular tuber crop and is grown as a vegetable in many parts of India.

Chemical Composition

Tender stem and leaves are also used for vegetable purpose. Tubers contain 18.0% starch, 1-5% protein and upto 2% fat. Leaves contain 2-3% protein, 3% carbohydrates and 4-7% crude fibre. Tubers and leaves are quite acrid due to high content of oxalates.

Climate and Soil

Amorphophallus is a tropical/sub-tropical crop and hence thrives well under warm humid climate with a mean annual temperature of 30-35^{0}C and a well-distributed rainfall of 1000 -1500 mm spread over a period of 6–8 months. The crop performs well under irrigated condition and hence could be grown even in areas where the rainfall is scanty but has assured irrigation.

Well-distributed rainfall or ample moisture availability during the growth phase of the crop is essential as it promotes proper vegetative development of the crop. It grows well on a variety of soils but a well drained sandy loam or sandy clay loam soil with a near neutral soil reaction (pH 6 -7) is ideally suited for the crop. The soil should be rich in organic matter with adequate amount of available plant nutrients.

Varieties and Hybrids

Gajendra and Santragachi are the two important non-acrid, high yielding Indian varieties of Amorphophallus.

Cultivation Practices

For minimizing the planting material requirement mini-sett technique has been developed for amorphophallus. In this case small corm pieces of 100 g weight are used for planting. Since the buds are located in a ring at the centre of the corm, setts are made out in such a way that each sett has a piece of the central bud. The setts are treated with cow dung slurry and dried under shade for one or two days before planting preferably in a nursery from where it can be transplanted to the field.

When the normal corm pieces are used as planting material then the spacing of 90 x 90 cm is adopted thus giving a plant population of 12,345 plants/ha while a reduced spacing of 60 x 45 cm gives a population of 37,000 plants/ha when the setts are used.

A good soil turning plough during early spring followed by pit formation is the traditional method of land preparation for *Amorphophallus*. The pit size should be of 60x60x45 cm. The top soil dug out is then mixed with farm yard manure or compost (2 -2.5 kg/pit) and the mixture is put back in to the pit prior to placing planting material over it. The planting material is placed vertically in the pits and is then covered with soil and compacted lightly. Amorphophallus is planted shallow, as deep planting would interfere with harvest operations, and more over, most of its feeder roots are found on the surface.

Mode of Pollination

It is a cross pollinated crop and dioecious in nature. Flowering is erratic, generally there is little or marginal overlapping of male and female flowering and female accessions flower earlier than male accessions. Flowering is observed as a rare phenomenon in this crop and generally occurs between May and June. Male accessions show consistency in flowering. Male flowers are produced in terminal and axillary panicles. Female flowers are borne on long axillary spikes,

1-2 spikes arising per axil. Pollinations are carried out as and when the corms produce spadics, which is always a single spadix per corm. The flowers are extremely protogynous, stigma receptivity lasting only 24 to 78 hours before spathe opening. Anther dehiscence takes place only 24 hours after spathe opening thereby preventing self-pollination. The opened flowers emit nauseating smell attracting carrion flies. They lay eggs upon the bulbous, crimson coloured appendix from which maggots hatch out and fly. Their role in pollination is not known. Fruit or seed set is not observed under natural condition. Pollen is yellowish in colour and sticky with more than 90 per cent fertility. Pollination conducted between 24 to 78 hours before anthesis were found fully successful resulting in fruit and seed set. Reception period of stigma could also be identified by the presence of sticky exudates over this surface.

Management of water, nutrients and weeds

Irrigation: Light irrigation soon after planting is recommended for uniform sprouting. Subsequent irrigations may be scheduled as per need. It is advisable to ensure that water does not stagnate in the field.

Nutrients: Well-decomposed farmyard manure is recommended to be applied @ 25-40 t/ha. Generally a fertilizer dose of 150:60:150 kg/ha of N P_2O_5 and K_2O is sufficient in most areas where it is grown. At the time of planting, the full dose of P and half N and K re to be applied in the pits. The remaining dose of N and K is applied 35 and 75 days after planting at the time of weeding and earthing up.

Mulching

Mulching is perhaps one of the most important cultural operations in EFY. It not only conserves soil moisture and temperature but also suppresses weed growth. It could be done either green or dried leaves.

Weed management: If proper mulching is done at planting, weeds would be suppressed to a large extent. Despite mulching two times manual weeding should be done in 45 days after the plantation and second weeding a month later the first weeding.

Harvesting and Yield

The crop is ready for harvesting by about 7-8 months after planting. Maturity is indicated by yellowing and drooping of the leaves.

Post harvest Handling

Care should be taken to avoid injury to the tubers while harvesting. After harvesting the tubers should be spread in shade for 3-4 days for curing. The soil and roots adhering to the tubers should be removed totally before disposal. Baskets or jute bags may be used for packing during transportation.

Diseases and Pests

The pests of this crop are similar to other aroids and yams as described under yam.

CHAPTER 13

Horse Radish (*Amoracia rusticana*)

Chromosome Number: 2n=4x=32

Horseradish, or *Amoracia rusticana,* is a member of the Cruciferae (mustard) family, belonging to the *Brassica campestris* group.

Origin

The horseradish has been around for a long time. The Romans referred to it as *amoracia.* It is native to Europe from Finland to the Caspian Sea. The "horse" in horseradish is used in the sense of "large," "strong," or "coarse"-thus the horseradish is in fact the coarse radish. Planted in early colonial gardens, it escaped and flourished as it does wherever it has been grown.

The name "horseradish" is thought to have come from an English adaptation of its German name. Germans called the plant "meerrettich" (meaning "sea radish") because it grew wild in European coastal areas. The German word meer (sea) sounds like "mare" in English. Perhaps "mareradish" became "horseradish". The word "horseradish" first appeared in print in 1597 in John Gerarde's English herbal on medicinal plants.

Botany

According to Shoemaker (1953) there are two types of horseradish the common and the Bohemian. The common form, which is broad leaved and hollow petioled, constitutes the most widely grown type. The roots are corky tan on the surface and white within. The Bohemian is a smooth, narrow-leaved, solid-petioled type. It lacks root quality and yield but is resistant to white rust and mosaic.

The plant is a hardy perennial and forms a white cylindrical tap root, which branches profusely at its terminal. The branches, which are of equal size, give rise to many lateral roots in the upper soil area. Some of the roots penetrate deeply into the soil. The plant grows up to 1 m in height with long, branching petioles. The lower leaves are from 15 to 30 cm long while the upper leaves are much shorter and more lanceolate. Since horseradish rarely produces seed under cultivation, the plant is propagated by root cuttings using either slender side roots or pieces of the main root.

Consumption and Uses

The horseradish root is peeled, washed, grated, and then mixed with a white or distilled vinegar solution (cider vinegar turns the grated horseradish dark). Since the primary use of horseradish is as a relish, recipes for preparing it commercially are kept a secret. Some blends use whipped cream; others sour cream and still others yogurt. The relish is a traditional accompaniment to roast beef and other meats. Noted for its sharp, nippy taste, horseradish can shares a place with other members of the mustard family at dinner table.

Major Cultivars

There is no any cultivar of horseradish in India. Other than the two basic types of horseradish there are no known cultivars. Because this vegetable sets no seed, it must be vegetatively propagated. Like the few other vegetables propagated in a similar manner, horseradish is not to be found in major seed catalogues. In establishing a new planting of horseradish, care should be taken in selecting good healthy stock from a strain that has a proven record of performance in the same general area.

Geographic/Climatic Requirements

The horseradish is a hardy perennial capable of being established under many different temperatures and growing conditions. Horseradish is grown in temperate climates around the world, primarily as an annual, and to a limited extent, as a perennial crop. It has long been an important crop in parts of Europe and significant production exists in the U.S. and Canada. When grown as an annual, it requires a long growing season. It needs warm temperatures during the summer growing season and cooler temperatures in the late summer and fall to enhance root development. In the annual system, the crop is usually planted in the early spring, and the entire root mass is harvested after the first killing frost in the fall.

Soil Requirements

Horseradish seems to grow anywhere, but does best in deep, rather moist soil. Moist sandy loam is suitable as are deep alluvial soils. The desired end product has long straight roots. Heavy or shallow soils may cause rough, pronged roots, which are wasteful in preparation for use.

Deep ploughing prior to planting, together with deep incorporation of fertilizers, is very beneficial. Soil pH on either side of 7.0 (6.0 to 7.5) seems adequate.

Fertilizer Requirements

To insure a fertilizer regime appropriate to the field and the crop, a soil analysis should be completed before attempting to fertilize horseradish. Due to the depth of the root system, deep incorporation below the root cutting planting zone is essential. If the soil contains too much rotted manure or too much N fertilizer is applied, horseradish will grow leaves at the expense of the root system.

Propagation and Cultural Practices

When the horseradish is grown as an annual, the roots are harvested in the late fall (the best time for the building up of reserves stored in the roots), the best roots are separated from the smaller roots and consumed. What remains are smaller roots and lateral roots. These are cut into pieces, and pieces with stem buds on them are retained. The side shoots are cut for replanting as long as they come from close to the main root. Planting can be in the fall before freeze-up or early in the spring after the frost is gone from the ground. Usually the root cuttings are planted in furrows 10 to 15 cm deep. Care must be taken to plant each cutting with the thickest end uppermost and about 25 to 50 cm apart in the row, with 67 to 75 cm between the rows. If left in the field for more than one season horseradish quickly becomes established and is increasingly difficult to control, and plants left for two years are woody and inferior in quality. For best results the roots should be lifted each fall. Although the roots can be harvested as soon as they are large enough, it is best to harvest only once-in the late fall. About 9 to 11 thousand root cuttings per acre are required.

Irrigation and Drainage

Horseradish remains underdeveloped with small roots if exposed to periods of drought; too much water is equally destructive. A well drained, moist planting bed is ideal.

Harvesting and Storage

The complete root system is deeply undercut and lifted so as to get most of the main roots. If any are left behind, they give rise to volunteer plants the second year-these can be treated with a contact herbicide. Side roots are trimmed off, and the large roots are packed in pallet boxes and may be bunched in six- or eight-plant lots if required.

Horseradish should be held at O°C and RH of 90-95 percent to prevent drying out. If exposed to light the roots become green. If free from diseases, the roots are capable of storing well for ten to twelve months.

Packaging

Horseradish is commonly packaged in 60-lb sacks, 50-lb sacks, 5-lb cello packages, or delivered in bulk for processing.

Pest Control

The same pests that attack other crucifer crops also attack horseradish. In using pest control, it is important to determine in advance the recommended regional practices from the local extension. In order to avoid a buildup of wilt and root rots, rigid rotation of horseradish plantings is necessary.

Disease control

Proper rotations, field selection, sanitation, spacing, fertilizer and irrigation practices can reduce the risk of many diseases. Fields can be tested for presence of harmful nematodes. Using seed from reputable seed sources reduces risk from seed-borne diseases.

Horseradish *Verticillium*

Cause: *Verticillium sp.*, a fungus that can survive for long periods of time as microsclerotia.

Symptoms: Unilateral wilting and death of above ground portions of the plant. Sometimes wilt is proceeded or accompanied by interveinal yellowing of the lower leaves of the plant and stunting of the plant. The vascular tissue of the root becomes brown or black in color, reducing the grade of the product when harvested.

Control: Long rotation is the only control available.

Horesradish white rust

Cause: *Albugo candida*, a fungus. Horseradish and radish are the principal economic hosts. The disease is known throughout the world where crucifers are grown. Many weeds are included in the host range. The fungus overwinters as oospores or mycelium in infected plant parts.

Symptoms: The fungus may attack any part of the plant: stem, leaves, or flowers. Systemic infections result in enlarged and distorted infected parts. Infected flower parts often lose their normal pigmentation and produce chlorophyll. The raised, white, spore-containing pustules, isolated or grouped together to form large patches, are the diagnostic symptom. The pustules have a powdery consistency and are produced on the leaves or stems. The fungus overwinters as oospores in plant debris or as mycelium in perennial plants. Leaf infection occurs readily at temperatures of 68 F when free water is present. Long periods of free water inhibit spore germination since spores require drying before germination occurs. Secondary infections occur by conidia that are windborne. Invasion of the plant is by germ tubes through stomata.

Control:

1. Destruction of horseradish debris aids in control.

2. Plant only roots from uninfected fields.

Bacterial leaf spot

Symptoms: Bacterial leaf spot of horseradish in some seasons rivals white rust in destructiveness, although it is not well known and has only recently been described. It is probable that this disease has been confused in the past with fungus leaf spots, especially by growers. The first evidence of infection is the appearance of small dark-green translucent spots which later enlarge and turn black. The spots are usually scattered over the entire leaf blade between the veins, but may be localized at places where water has been retained longer than usual. They are especially conspicuous on the under surface of the leaf. The spots rarely exceed 1/4 inch in diameter, and vary in outline from definitely circular to angular. The lesions may remain quite small during periods of dry weather and then enlarge considerably after rains. The killed tissues fade with age so that later the central area may appear light brown in color, though ordinarily the spots are much darker than those produced by other organisms.

Control: No experiments on the control of this disease have been reported. It appears from our meager knowledge of the habits of the organism that sanitation yields the best results. This consists of as thorough elimination of old leaves and crowns as possible.

Cercospora leaf spot of horseradish

Symptoms: The young spots are bright colored, but as the lesions progress, they turn dark because of the presence of fruiting bodies of the causal fungus, *Cercospora armoraciae* Sacc. The spores themselves are hyaline, but the conidiophores are brownish black. Note the concentric ringing of the older spots in gray color. Later the spots may appear almost black, due to the presence of the numerous conidiophores of the fungus. Sometimes these appear in definite concentric rings, but this is not generally the case. The spots often coalesce, forming large dead areas covering the greater portion of the leaf. Leaves having a number of lesions are easily scorched by the hot sun and the dead areas may extend beyond the limit of the infection or the whole leaf may wither. Little is known concerning the time and method of infection, but judging from our knowledge of similar diseases, it is probable that the mycelium of the fungus overwinters in old leaves and gives rise to spores the following spring. Alternate host plants may also be responsible for spring inoculums.

Control: No method for the control of this leaf spot based on data from experiments has been reported. It has not been considered serious enough to warrant special control methods.

Ramularia leaf spot

Cause: Ramularia leaf spot is caused by *Ramularia armoraciae* Fuckl.

Symptoms: Ramularia leaf spot has been reported on horseradish from practically all countries where this crop is produced. It seems to have a wider distribution than *Cercospora* spot. Reports from Poland and Russia, together with those from all sections of the United States and Canada, indicate that this leaf spot is cosmopolitan. It appears to be the predominating leaf spot of the western United States. Ramularia leaf spot is quite similar in appearance to the Cercospora leaf spot with which it is often confused. Lesions are found on the leaf blade, usually between the larger veins, appearing first as yellowish-green circular areas. Later the spot becomes more definite, and there is often a separation of the thin, papery, diseased area from the healthy tissue. This area may drop out, giving a shot-hole effect. The lesions are frequently irregular in out- line due to an uneven advance of the fungus from the central area. Later in the season dark bodies resembling pycnidia may be observed over the surface of the lesion. Withering and drying of the entire leaf results when the infection is heavy.

Control

There have been no reports of the disease becoming serious enough to warrant special control measures; accordingly, no information is available regarding control methods.

CHAPTER 14

Ivy gourd (*Coccinia grandis* (L.) Voigt)

Chromosome number: $2n = 24$

Origin:

Ivy gourd (*Coccinia grandis*) occurs wild from Senegal east to Somalia and south to Tanzania; also in Saudi Arabia, Yemen and India. It is locally naturalized in Mozambique and Mauritius and has been introduced in many other tropical and subtropical regions; it is considered an invasive, obnoxious weed in Australia, Florida and the Pacific Islands, e.g. Hawaii. It is cultivated in Asia from India to Indonesia. In East Africa, especially in Kenya, it is cultivated mainly for consumers of Indian origin.

Uses:

Ripe red fruit of sweet-tasting cultivars are eaten raw, or they are peeled and cut into pieces and prepared as a stew with onions and tomatoes. Other vegetables may be added such as garden egg, bottle gourd, ridge gourd, okra and pulses. Meat is added when available. The dish may be served with bread, sorghum or any other starchy food. The dish is known in East Africa as 'tindori'. Immature green fruits are prepared in soups and curries and widely used in this way in Ethiopia and India. Cultivars with bitter fruits are mainly used for their leaves and shoots. Such shoots are especially popular in Thailand where they are eaten as a fried or boiled vegetable. The leaves are also eaten as a vegetable by the Mursi people in Ethiopia. The seeds are chewed in Ethiopia and Kenya, but they are discarded in West Africa and Uganda. The fruits, stems and leaves have medicinal uses such as to reduce high blood pressure and to treat abscesses. Roots are believed to heal illnesses associated with endocrine system disorders such as diabetes mellitus, and are used in Niger to treat intestinal troubles. Bronchial inflammation, respiratory mucosae and skin disorders are also said to be healed by this plant. In some communities in India ivy gourd is cultivated as a medicinal plant. Goats, sheep, cattle, donkeys and camels eat the plant.

Properties:

Per 100 g edible portion, the fruits contain: water 93.5 g, energy 75 kJ (18 kcal), protein 1.2 g, fat 0.1 g, carbohydrate 3.1 g, fibre 1.6 g, Ca 40 mg, P 30 mg, Fe 1.4 mg, thiamin 0.07 mg, riboflavin 0.08 mg, niacin 0.7 mg, ascorbic acid 1.4 mg (Rubatzky, V.E. & Yamaguchi, M., 1997). Accurate data on the composition of the leaves are not available.

Description

Dioecious perennial herb up to 20(–30) m long, trailing or climbing with simple tendrils, with tuberous rootstock; stem green and longitudinally ribbed when young, becoming white-spotted when older and eventually woody. Leaves alternate, simple; stipules absent; petiole 1–5 cm long; blade broadly ovate to pentagonal or orbicular in outline, 3–12 cm × 3–15 cm, shallowly to deeply palmately 3–5-lobed, cordate at base, margin entire or sinuate and often with distinct reddish glandular teeth, glabrous, punctate. Flowers axillary, unisexual, 5-merous, with tubular receptacle 3–7 mm long, sepals linear, up to 6 mm long, corolla campanulate, with lobes up to 2 cm × 1.5 cm, yellow-orange; male flowers solitary or paired, rarely 3–4 in a short raceme, pedicel 1–7 cm long, stamens 3, united into a column; female flowers solitary, pedicel up to 2.5 cm long, ovary inferior, cylindrical, up to 1.5 cm long, 1-celled, style 3 mm long, stigma 3 -lobed. Fruit an ellipsoid or rarely spherical berry 3–7 cm × 1–3.5 cm, fleshy, green with white stripes when young, turning red at maturity, many-seeded. Seeds asymmetrically pear-shaped in outline, compressed, 6 mm × 3 mm, margin rather thick and grooved.

Other botanical information

Coccinia is placed in the tribe *Benincaseae* and comprises about 30 species, all of them confined to tropical Africa with the exception of *Coccinia grandis*. Anchote (*Coccinia abyssinica* (Lam.) Cogn.) is cultivated mainly for its tuberous fleshy rootstock in Ethiopia, but its young shoots are eaten as a cooked vegetable. In Malawi children may eat the fruits of *Coccinia rehmannii* Cogn., although they are said to cause sore eyes; the starchy tuber is eaten after roasting. The leaves of *Coccinia trilobata* (Cogn.) C.Jeffrey are a famine food, occasionally used as a relish by some tribes in Kenya, e.g. in the Mbeere district. The fruits are said to be toxic.

Growth and development:

Cuttings start developing new leaves and shoots after about one month. Suckers develop after another month, and flowering and fruiting start again one month later. Ivy gourd plants flower profusely and produce many fruits. Flowers only develop on the branches and not on the main stem. Young fruits can already be harvested about one week after flowering. The period from planting to the first harvest takes up to 5 months. Harvesting is best done weekly as long as there is enough moisture to produce new leaves and flowers. The crop could remain in the field for up to 10 years, after which the yields decline and farmers need to establish a new crop. However, replanting after 4 years may be more economical. In the wild, plants can become 20 years or older and the stems may become 30 m long with a diameter of up to 7 cm at the base. Stems become corky with age and both the rootstock and stems act as water reservoirs that enable the plant to survive the dry season.

Cultivation of Ivy Gourd

There are few high yielding varieties of ivy gourds which are available in the market. For obtaining a good yield, a good yield must be selected. The cultivation of ivy gourd must also be done properly. The plant is invasive and can form a weed very easily. So, it must be cultivated properly. The modern methods of cultivation must be followed to get good quality of fruits.

Soil Conditions

The ivy gourd grows well in well-drained soils. The soil must be fertile and rich in organic matter. The ivy gourd can grow in other types of soils also. Good soil must be used for ivy gourd cultivation. Only then, the production will also be good. The soil must be well aerated. The ph of the soil must not be extremely high or low. A neutral ph of the soil is preferable for farming. The soil should not be too deep or light.

Climatic Conditions

The ivy gourd is basically a crop of the tropical climate. The crop must not be grown at too high altitudes. Full sunlight is needed for the proper growth of the plants. It can also be grown in partial sunlight. In either case, the crop must not be exposed to extreme climatic conditions. In winter, frost and extreme cold must be avoided. Too much rainfall is harmful to the crop. A moderate rainfall is sufficient for the ivy gourd cultivation.

Land Preparation

The land must be thoroughly cleaned and made free of weeds and other materials. Once the desired land is selected and cleared, it must be made suitable for farming. After that, the field must be ploughed several times. Ploughing makes the soil texture smooth. After ploughing, the land must be levelled and harrowed. Farmyard manure must be mixed with the soil after the last plough. Drainage channels must be made in the field. The drainage channels must be made so that extra water can be drained out of the field.

Propagation Methods

The ivy gourd is propagated through seeds. The seeds must be selected properly. Only healthy seeds which are disease free must be selected for sowing. The seeds must be pre-treated before sowing them in the soil. Pretreatment prevents the fungal and other such diseases. It is better if the seeds are bought from a reliable source.

The seed propagation takes a longer time to germinate and mature. Therefore, ivy gourd is also propagated through vegetative methods such as by cuttings.

Planting Methods for Ivy Gourd Farming

The cuttings are raised in nurseries. When they are suitable for planting, they can be transplanted to the main field. Healthy plants and cuttings must be used for vegetative propagation. For transplanting, pits must be dug in the main field. The pits must be exposed to the sunlight for about a month. After that, it must be filled with farmyard manure and topsoil. The cuttings must be planted in the pits and watered immediately.

The seeds can be directly planted in the main field or can be transplanted to the main field from the nurseries. The seeds usually take 3-4 weeks to germinate. In nurseries, they are grown under controlled conditions. After the seedlings have developed roots and few leaves, they can be transplanted to the main field. They must be planted carefully in the main field and watered. Only healthy seedlings must be selected for planting in the main field. The seeds can be directly sown in the field. any seed sowing methods can be used for sowing. The seeds must not be planted too deep in the soil. any seed sowing equipment may be used for sowing of the seeds.

While sowing, proper spacing must be given. There is no specific planting season for ivy gourd. The ivy gourd can be planted any time of the year. The plants must be raised in the greenhouse when young. This must be done so that the young plants can be protected against harsh environmental conditions.

Irrigation

The ivy gourd plants need more water while it is still young. The first irrigation must be provided when the sowing is done. After that, irrigation must be given based on the moisture content of the soil. in times of extremely dry and hot weather, proper irrigation must be provided. To save water, sprinkler and drip irrigation can be used. In rainy weather, irrigation is not required. If there is too much rainfall, the water must be drained out of the field. For that, drainage channels must be provided in the field. In winter, the irrigation must be given based on the soil moisture. Little irrigation is needed to protect the plants from frost. Watering the plants is extremely important during fruit formation and flowering time.

Fertilizer and Manure Application

A soil test must be done before application of fertilizers. Based on the results, the fertilizers must be applied. Farmyard manure must be applied at the time of land preparation. Too much chemical fertilizers must not be applied. It makes the soil dry and barren. The organic fertilizers must be applied along with chemical fertilizers in the land. After each time fertilizer is applied, a light irrigation must be given. Application of nitrogen, potassium, and phosphorus in the right amounts is very important. To avoid micronutrient deficiency, a proper quantity of micronutrients must also be applied.

Intercultural Operations

The ivy gourd itself can become a weed problem if crop management is not done properly. The weeds from the ivy gourd crop must be removed at an early stage. The weeds can be removed by hand. Chemical pesticides can also be used in case of extreme problems. Sometimes, cultural methods also work better. A combination of all the methods can also be useful.

The ivy gourd is a vine plant. When the plants start growing, they must be trained on a trellis or provided a suitable support system. The training of the plants is important. Pruning can be done whenever necessary.

Harvesting

Farmers harvest twice weekly.

Yield

Individual plant yields are in the order of 7–10 kg of immature fruits per year; per ha 10–13 t/year may be harvested. Peak production is between the second and fourth year, after which the production declines unless the crop is regularly fertilized, irrigated and pruned. There are no data on yields of leaves.

Diseases and pests

Ivy gourd is host to a range of diseases including powdery mildew (*Sphaerotheca fuliginea, Erysiphe cichoracearum*). Seed-borne anthracnose (*Colletotrichum lagenarium*) may cause heavy losses of fruit especially in storage; it can be controlled by spraying with copper oxychloride. Other diseases include black rot (*Curvularia pallescens*) and stem rot (*Rhizoctonia solani*). Viral diseases, such as watermelon mosaic virus (WMV), may affect the plant. The symptoms include mottling, green blisters, curled leaf margins and distortion of fruit. A major pest in cultivated crops in Kenya is caused by coreid bugs that feed on new shoots and young fruits. When sucking fruits, they cause distortion of fruits or fruit shedding, whereas a heavy infestation can cause dieback of the shoots. The pumpkin caterpillar (*Diaphania indica*) can be a serious pest on ivy gourd. Aphids (*Myzus persicae, Aphis gossypii*) feed on the flowers and affect the development of new fruit. Fruit flies can also be a serious pest. Pesticides should only be applied when needed because they might also kill beneficial insects. Other pests found in Kenya are a weevil (*Acythopeus cocciniae*) and clearwing moth (*Melittia oedipus*) of which the larvae are stem borers that cause the stems to collapse. These two natural enemies are now used as biological control in areas where ivy gourd is a noxious weed.

Handling after harvest

The fruits are spread out during the evening in a single or double layer on top of clean sacks. When they are covered or kept inside a sack, they tend to turn yellow outside and pink to reddish inside. Only fresh green fruits which are whitish to pale green inside can be sold. Traders use cold rooms to keep them fresh. When stored in a refrigerator, the fruits last for up to two weeks. Prior to taking the fruits to the market, they are placed in cartons, each containing about 7.5 kg. These cartons are also used for exporting the fruit.

CHAPTER 15

Jerusalem artichoke (*Helianthus tuberosus* L.)

Family: Compositae

Chromosome number: 2n= 102

Jerusalem artichoke, or *Helianthus tuberosus* L., is a member of the Compositae (composite) family. It is also known as the sunchoke, girasole, and sunroot. It cultivated for its oblong tubers. It is native of North America. The cultivation of this crop has not expanded. It is recommended as feed for hogs and also used as a human food. The carbohydrate insulin is stored in the tubers. This is of interest because of it's value in the diet of persons suffering from diabetes. It has wide adaptability to a range of soil and climatic condition. It is not yet popular in India.

Origin

The Jerusalem artichoke is a North American native that grows wild throughout large areas of the United States and Canada. The unrelated name "Jerusalem" was appended through confusion with the Italian *girasole*. Europeans first became aware of this species in the seventeenth century, when it became noted for surviving under conditions too rugged for the white potato.

Botany

A sunflower like herbaceous perennial, the plant forms knobby potato like tubers ranging in color from white to yellow and from red to blue. The tubers are unusual in that they do not form starch but rather insulin, a polymer of fructose that upon hydrolysis yields fructose (levulose), which is of value to diabetics. The insulin content is at its peak after a killing frost has occurred.

The Jerusalem artichoke often is a tall, erect plant growing to and beyond 3 m, branching in the lower parts, with pointed leaves very rough to the touch. The flowers are small, the florets yellow. The tubers form late in the growing cycle and should not be dug until the stems are frozen. The flesh is sweet and somewhat watery. If left in the soil, the tubers are not damaged by winter freezing.

Consumption and Uses

The Jerusalem artichoke is delicious raw and should never be overcooked. Lightly cooked tubers are served in salads. Tubers have a texture and flavor similar to lychee nuts (*Litchie Chinensis*) in Chinese cuisine. In oriental cooking they are added to soups and may be pickled. The key production states are California, Michigan, and New Jersey. There is no definable production area in Canada. As a new crop in the biotechnology area the Jerusalem artichoke may become one of the new wonder crops due to its high potential for developing insulin.

Major Cultivars

There are no recently introduced cultivars. Mammoth White French and Sutton's White are two old cultivars. The produce industry usually refers to Jerusalem artichokes as sunchokes.

Geographical Climatic Requirements

The Jerusalem artichoke is well adapted to almost any geographic situation. In most places it establishes itself quickly and can become a weed plant. Prior to development of effective herbicides, this was a problem. Like the horseradish, the sunchoke establishes a new plant with every tuber left in the soil. The plant adapts well to almost any condition provided the soil is well drained. For maximum yields a planting bed should provide good drainage and adequate moisture and nutrients. Where maximum or optimum growing conditions are not present, this plant manages to produce a crop even though it may be light. Similarly, mean temperatures are satisfactory over much of the United States and Canada.

Cultural Practices

Thompson and Kelly (1957) reported German and French studies indicating that K is important in the growth of Jerusalem artichoke. Fertilizer requirements for white potatoes would appear to give excellent results. Like the white potato, the Jerusalem artichoke is planted using tubers of about 50-60-g size. The need for cultivation for weed control is very low due to the rapid growth of the plant. For a full crop stand, spacing is usually such that only one-half the seed (tubers) required for white potato is necessary.

Pest Control

For adequate control of pests, local recommendations should be followed.

Weeds: An unusual crop plant, the sunchoke seems able to compete successfully with weeds under most conditions. When the plant itself becomes a weed, there are a number of effective herbicides that can be used to eliminate unwanted proliferation.

Insects: Aphids and nematodes are known to infest this crop but rarely do they become a serious problem.

Diseases: The Jerusalem artichoke is subject to various storage and soft rots after the tubers have been harvested. Again local recommendations should be followed in correcting this problem when detected.

Harvest and Storage

Before the tubers can be harvested, the tops must be removed. If provision has not been made to contain the root system, it may be difficult to harvest the tubers without doing considerable damage. If converted potato diggers cannot successfully lift the tubers, they must be carefully dug by hand, therefore every effort is made to machine dig and lift the tubers. Exceptional care must be taken to prevent damage to the very sensitive thin tuber epidermis. Any skin damage causes rapid loss of moisture with resulting shriveling of the tubers and greatly reduces the normal storage-four to five months when held at about O°C in a high humidity environment. Without temperature and humidity control, the tubers are not likely to last more than a few weeks. Storage rots are quick to set in if careless handling occurs during harvest.

CHAPTER 16

Kachari (*Cucumis callosus*)

Family: Cucurbitacae

Chromosome number: 2n= 14

In Sanskrit, it is also known as chitra, chitraphala, mirgakshi whereas in engllish it is regarded as cucumber pubescent. It is very hardy plant, monoecious in nature and highly cross- pollinated. Anthesis starts between 5 am to 9 am. Fruit is small, egg shaped; sub globose, 3 to 125 in weight. The skin colour varies completely white to dark green and it's various shades. At immature stage, it is generally green and bitter in taste but turns to pale yellow and sweet at ripening. Mature fruits are consumed in various forms as cooked vegetables, chutney and pickles. Powdered typc of kachari is uscd in various types of spices on commercial scale. The dehydrated kachari is diuretic, gastric- stimulant, cooling refrigerant, coughicide and wormicide. It is useful in digestion, facial paralysis, debility, bile, obesity, piles, constipation troubles and cough.

Climate and soil

It can be grown under warm climate of arid and semi- arid conditions. The optimum temperature for germination of seeds is 25°C whereas for growth and fruiting it needs 30- 35°C. Heavy rainfall area is not suitable for its cultivation because its vine get damaged due to foliar diseases. It can be grown on any type of soil but a well drained, sandy soil is ideal for its cultivation.

Varieties:

Some work has been initiated at CIAH, Bikaner and two varieties have been developed which characteristics are given below:

AHK- 119: it is an early, high yielding selection from the land races of kachri. The fruits are suitable for processing. Fruits are small, egg shaped, 50- 60 g weight, 5-6 cm length and 4- 4.5 cm diameter. It can be sown in February- March as an irrigated summer crop and in July as a rainfed crop under the arid condition.

AHK- 200: it is very early maturing variety and possess drought tolerance characteristics. It has been selected from the land races growing in the arid region. Fruits are 100- 120 g in weight, 7.5 cm in length and 5.1 cm in diameter. The fruits are ready for harvesting 65- 67 days after sowing and continue to fruit for 90- 100 days. On an average, 20 fruits can be harvested from single vine giving a yield of 115- 120 q/ ha. The variety is suitable for sowing in summer as well as in rainy season under arid season under arid environment.

Cultivation:

It is propagated by seed and naturally germinates before rain. However, it can be grown during summer and rainy season also. In summer, sowing is done in the month of February- March, whereas, June- July during rainy season. In rainy season it is cultivated as mixed crop with *jowar* or maize. As a pure crop, sowing in bed and pit system is generally followed. Usually sowing is done on top of furrows at 1.5- 2.0 m apart and vines are allowed to trail on the bed especially in summer. The hills should be made at a distance of 60 cm. About 2-4 seeds are sown in a pit and finally two vines are retained. About 2.0 to 2.5 kg seeds is enough for sowing one hectare crop.

Manures and fertilizer:

Generally well rotten FYM @ 20- 25 tones, besides 80 kg N, 40 kg P_2O_4 and 40 kg K_2O/ha is applied at the time of land preparation. Apply 1/3 N and full dose of phosphorus and potash as basal dressing. Remaining dose of nitrogen should be applied 30- 45 days after sowing at top dressing.

Irrigation:

In summer season sufficient moisture especially during maturity period is essential. Summer season crops require 5- 6 light irrigations, whereas, rainy season crops need only life saving irrigation during the period of long dry spell.

Harvesting and yield:

The fruits are harvested at maturity stage. At immature stage fruits are bitter in taste that gradually becomes sweet and sour towards maturity. The fruits attain maturity at 65- 75 days after sowing. Mature fruits have very long storage life; so can be easily transported to distant markets. The yield varies from 100- 125 q per hectare.

CHAPTER 17

Khohlrabi (*Brassica oleracea* L. var. gongylodes)

Family: Brassicacae

Chromosome Number: 2n= 18

Origin

Kohlrabi shares with Brussels sprout the distinction of originating in northern Europe approximately 500 years ago. It is speculated that it was derived from the marrow or wild cabbage. The marrow cabbage is a cold temperature sensitive cole crop that develops very thick heavy stems. The first recorded evidence of kohlrabi was in 1573 in what is now Germany, where it was particularly well known. It is still very important there as well as in England and the eastern Mediterranean countries. The first recorded mention of it in the United States was in 1806.

Botany

Kohlrabi is an ancient member of the *B. oleracea* or cabbage group. A herbaceous biennial, the plant develops a thickened stem above the soil level. The swollen part is a spherical body with leaf stems initiating at random from broad leaf scars on its surface. Like the spherical body, the leaf petioles are fleshy and succulent. The leaves are smooth and covered with a pronounced bloom. Initially a tap root forms and quickly develops numerous branches of about equal size.

Kohlrabi is highly temperature sensitive and will vernalize after exposure to prolonged temperature below 100C at any stage of development. Under the conditions brought about by vernalization, kohlrabi becomes an annual. There are two types of tubers based on color white spheres or stems and the purple type.

Consumption and Use

There is not a large demand for kohlrabi in North America. A few thousand hectares or so are grown. In West Germany, where this crop is popular, only about 1600 ha are in production. Usually the hectares per farm planted to kohlrabi are quite limited.

Kohlrabi is used raw or cooked (steamed) and the tender leaves are used as greens. Cooked kohlrabi can be scooped out, stuffed with onions and sausage, and baked. Kohlrabi can also be served with a cream sauce or cheese sauce.

Nutritive Value: Knol-knol (*Brassica oleracea.* Gongylodes group), fresh, raw, Nutrition value per 100 g. (Source: USDA National Nutrient data base;

Principle	Nutritive Value	Electrolytes	
Energy	27 Kcal	Sodium	20 mg
Carbohydrates	6.20 g	Potassium	350 mg
Protein	1.70 g	**Minerals**	
Total Fat	0.10 g	Calcium	24 mg
Cholesterol	0 mg	Copper	0.129 mg
Dietary Fiber	3.6 g	Iron	0.40 mg
Vitamins		Magnesium	19 mg
Folates	16 µg	Manganese	0.139 mg
Niacin	0.400 mg	Phosphorus	46 mg
Pantothenic acid	0.165 mg	Selenium	0.7 µg
Pyridoxine	0.150 mg	Zinc	0.03 mg
Riboflavin	0.020 mg	Calcium	24 mg
Thiamin	0.050 mg	**Phyto- nutrients**	
Vitamin A	36 IU	Carotene-ß	22 µg
Vitamin C	62 mg	Crypto-xanthin-ß	0 µg
Vitamin K	0.1 µg	Lutein-zeaxanthin	0 µg

Major Cultivars

Two main cultivars, Early White Vienna and Early Purple Vienna. Currently three F 1 hybrids are available-Grand Duke, White Danube, and Purple Danube.

Climatic Requirements and Production Regions

Kohlrabi shares with the other cole crops a need for cool temperatures. However, Kohlrabi is more sensitive to prolonged exposure (one week or less) to temperatures of l0oC or less, which will either bypass tuber production, or, if the period is less than one week, will initiate premature flowering after the tuber has formed. The sensitivity to low temperatures begins with the germinating seedling, which will upon exposure to temperatures of -1.1° to 1.1°C initiate vernalization. Other cole crops for the most part respond to vernalizing temperatures only after the plant is fully developed. Along with cabbage, collards, kale, and turnip, kohlrabi is considered hardy. The main commercial areas of kohlrabi production are Texas for the winter crop and Illinois and New Jersey for the summer crop. Home gardeners throughout the northern tier of North America grow kohlrabi. Sequential sowing progresses with the season to provide a steady flow of product year round. For maximum quality kohlrabi should be grown quickly without water stress. Slow growth has a tendency to make the tubers woody and cause them to crack. It is essential that a good supply of irrigation water be available should the normal rainfall be inadequate. As for all cole crops, good drainage is essential.

Pest Control

For kohlrabi and other cole crops good growing practices such as rotations and seed treatment will help reduce losses from pests. The essential weed, insect, and disease control methods are similar to those for all the cole crops. Local recommendations and licensing regulations should be followed. Kohlrabi is susceptible to black rot and bacterial soft rot, which are common storage and marketing diseases.

Harvesting:

Kohlrabi is hand harvested at the 5-7-cm stage. However, late maturity cultivars may attain 10- 12 cm diameter at harvesting stage. The appearance of the fresh green or purple root and healthy leaves is essential.

Yield:

Average yield may be 12- 20 t/ ha depending upon cultivars and other agro climatic condition.

Storage:

Storage of harvested tubers should be held at a temperature below 5°C and a relative humidity of about 95 percent in order to prevent excessive wilting and toughening of the flesh. Kohlrabi is topped and packaged in perforated film to reduce moisture loss. There are indications that kohlrabi shares with asparagus the tenderizing effect of CO_2 enrichment in storage. Held at O°C kohlrabi can be effectively stored for up to one month.

CHAPTER 18

Pigweed/ Bathua (*Chenopodium album* L.)

Family: Chenopodiaceae

Chromosome Number: 2n= 36, 54

Origin: It is native of Europe and Western Asia

Botanical Description

It is a fast-growing, plant, commonly found in field as weed. It tends to grow upright at first, reaching heights of 30-80 cm. Stems: Rarely slender, angled, often striped green, red or purple. Leaves: Simple, rhomboid, deltoid to lanceolate, upper entire, lower toothed or irregularly lobed, extremely variable in cultivated forms, 10-15 cm long, petioles often as long as thick blade, 1 to 1.3 cm in length. The opposite leaves can be very varied in appearance. The first leaves, near the base of the plant, are toothed and roughly diamond-shaped, 3-7 cm long and 3-6 cm broad. It has been found in dark green colour with smooth undersurface. The leaves are waxy-coated, unwettable and mealy in appearance, with a whitish coat on the underside. Flowers: Radial, symmetrical and grow in small cymes on a dense branched inflorescence, 10-40 cm long, contains shining black seeds. Its pollen can contribute to hay fever-like allergies.

Nutritive Value

The nutritional composition of *Chenopodium album* leaves per 100 g edible portion is: water 84 g, energy 184 kJ (44 kcal), protein 4.3 g, fat 0.8 g, carbohydrate 7.3 g, fibre 2.1 g, Ca 280 mg, P 81 mg, vitamin A 11,300 IU, thiamin 0.15 mg, riboflavin 0.4 mg, niacin 1.3 mg, ascorbic acid 90 mg (Rubatzky, V.E. & Yamaguchi, M., 1997). The nutritional composition of seed of Himalayan cultivars per 100 g is: energy 1654 kJ (395 kcal), protein 16 g, fat 7 g, carbohydrate 66 g (Partap, T., Joshi, B.D. & Galwey, N., 1998). Seed of the weedy types of *Chenopodium album* (in Africa and elsewhere), however, is probably of inferior quality and less nutritious. An ethanolic extract of the fruits has shown antipruritic and antinociceptive activities in tests with mice.

Cultivation

Grain chenopods are commonly sown 1–2 cm deep in rows 25–50 cm apart depending on soil moisture content and expected rainfall. The 1000-seed weight is about 1.4 g. Seed rate is 6–10 kg/ha, resulting in 100–150 plants per m^2. The seedbed should be well prepared. The seed may also be broadcast, but it is easier to weed when sown in rows. Broadcasting requires about 20 kg of seed per ha. Seedlings will emerge in approximately one week in a sufficiently moist soil with an average temperature above 10°C. The period between sowing and flowering and from flowering until maturity of the seed is very variable. Early and daylength-neutral cultivars may take 50–60 days to flowering and 90–110 days to seed maturity, whereas late and short-day cultivars need 4–5 weeks longer. Grain chenopods are predominantly self-pollinating; cross pollination is less than 10%. The most important disease is downy mildew (*Peronospora farinosa* f.sp. *chenopodii*), which causes much damage in chenopod growing areas all over the world. The disease is favoured by warm and humid weather. Some cultivars are partially resistant. Other fungal diseases may also cause serious damage, as well as insect pests, of which the leaf miner or leaf sticker *Eurisaca melanocompta* is the most serious. Birds attacking the crop before harvesting or during field drying probably cause the greatest crop losses. Bitter cultivars are less prone to such attacks than sweet ones.

Harvesting methods used for cereals can also be used for grain chenopods. Plants are cut, bundled and dried, threshed, and the seed is winnowed.

Yields of 0.2–0.6 t/ha are reported from farmers' fields in India. One reason for low yields may be that grain chenopods are not often grown alone but are usually intercropped. Inflorescences do not dry easily. At harvest, seed moisture content may be around 20%, and artificial drying to 14% moisture may be necessary.

CHAPTER 19

Rhubarb (*Rheum rhaponticum* L.)

Family: Polygonaceae

Chromosome number: 2n= 44

Rhubarb, or *Rheum rhaponticum* L., also *Rheum hybridicum* L. *R. rhabarbum* L, is a member of the Polygonaceae (buckwheat) family. The name comes from the Greek words *Rha,* the name for the river Volga, and *barbaron* meaning "barbarous" or "foreign"-thus a plant from the foreign Volga region.

Origin

The rhubarb was known in China about 2700 B.C., according to herbalists of Peking, as a medicinal rather than as a food plant. The Romans knew of the medicinal qualities of the dried rhubarb root. Marco Polo brought the rhubarb to Europe, where English monks cultivated it first as an herbal and later as a food plant. By the eighteenth century it had become established as a food plant. About this time European settlers brought rhubarb to America, where it spread quickly.

Botany

Rhubarb is a herbaceous perennial that begins its new growth each year from a semi woody crown topping large, thick, fleshy storage roots. Buds develop on the top of the crown beginning at the center and moving outward as the plant ages, leaving the center to dry out and die. In time the plant can spread outwards more than a meter leaving a large unproductive center. Buds always develop the year prior to producing shoots. Each bud may have as many as thirty differentiated leaves by the time dormancy sets in. Growth of each bud begins from the outside. As each leaf expands the sheath breaks and growth is rapid. Depending upon the cultivar, the elongated and thickened leaf petiole that is the edible portion of the plant usually grows, to a height of 75 cm and in some instances to a width of 6-7 cm at the base. The leaf blade is very large, sometimes growing to as much as 90 cm in length and breadth.

Seed stalk development occurs at different rates depending upon cultivar and spacing between plants. Close spacing has a tendency to restrict seed stalk formation. In a producing planting, the seed stalks are pulled up and destroyed as they appear in order to increase leaf development. Flowers are small and greenish white; they form fruits that have wings for wind dispersal of the seed.

Consumption and Uses

Rhubarb at one time was known as the pie plant. The petioles are chopped and used in pies and in mixes with other fruits to make preserves. The medicinal characteristics of rhubarb have not been forgotten and so this vegetable has a dietetic as well as culinary usefulness.

There are no precise figures on the amounts of rhubarb consumed in the United States. Canadian consumption has been estimated a around .14 kg-more than ten times that amount is probably being consumed in the United States.

Major Cultivars

Rhubarb is both an outdoor plant and a forcing plant. Cultivars for outdoor production are Timperley Early, Valentine, Macdonald, and German Wine. For forcing, Timperley Early, Victoria, Crimson Red, Strawberry, and Sutton are available. For very high-quality outdoor production, the red-petioled cultivars Ruby and Canada Red are preferred.

Geographic/Climatic Production Requirements

The rhubarb is a perennial plant widely adapted to nearly all regions. In the warmer southern regions, high temperatures disrupt growth. In the northern regions, the plant usually remains dormant in the hot dry summer period and becomes very active in wet or cooler periods. Whether grown at high or low elevations, rhubarb responds to temperatures between 4° and 24°C. Cultivars that exhibit good pink or red petiole color during the cool season usually become very green when temperatures remain above 30°C. With rhubarb grown for forcing in areas such as Oregon, Michigan, and Washington, the petioles remain green and vigorous during the summer and display refined pink color under the forcing conditions of minimum light and comparatively low temperatures. Throughout the northern United States and Canada rhubarb is a staple vegetable in home gardens.

Soil and Fertilizer Requirements

Rhubarb is grown in a wide range of soils. A well-drained loam containing a high degree of organic matter is the best type of soil for rhubarb. In addition rhubarb has a wide pH tolerance from moderately acid to somewhat alkaline. Since rhubarb is a heavy feeder, it does well planted on soils with a heavy incorporation of manure prior to the planting of the crop and a heavy application (determined by soil analysis) of inorganic fertilizer in the form of phosphoric acid and potassium chloride. The incorporation of manure prior to planting the rhubarb and subsequent working it in after planting will provide a good nutrient base throughout the life of the stand. On particularly rich clay or silt loam soil, lesser amounts of P and K may be satisfactory. According to Thompson and Kelly (1957) a 1: 1: 1 ratio with 400 kg per ha is usually adequate.

Cultural Practices

Rhubarb is planted by root crown divisions of two- or three-year-old plants. A normal division consists of three to eight new plants. Care should be exercised not to make each division too small. Whether for forcing or outdoor production, rhubarb is established from carefully selected parent plants in order to insure the best performing new plants. Since rhubarb seed is highly heterozygous propagation is the best method of establishing new plants. In field production about 12 to 13 thousand plants per ha are planted.

Rhubarb grown for forcing is usually kept separate from outdoor produced plantings, and as a rule the stalks or petioles are not harvested. Two years after field planting, the plants are ready to be dug for forcing. If left until they are three years old, they become too heavy and bulky to manage efficiently. High N fertilizer is applied at the beginning of the second season by drilling it into the side of the plant row at about 10-12 cm deep. Soil analysis should be carried out to determine if P and K are required as well. Rhubarb plants should not be permitted to set seed or actively develop flower stalks because of the heavy drain on the root system. In order to get the best response to forcing, the rhubarb plant must be exposed to a period of low temperatures to break dormancy once the forcing conditions are started under cover. The crown and roots must be dug before the soil freezes, taken to a sheltered storage, and held where dehydration and water loss are at a minimum until the forcing operation begins-usually early in the new year. When ready for forcing, the crowns are packed together on 11/2 m-wide beds-with soil and water to initiate growth. The optimum forcing temperature is around 10-11°C. Anything lower than this will enhance petiole color but also greatly retard shoot growth. Regular watering and near constant temperature in the forcing shed are the basic requirements. The first harvest occurs after about one month. Harvest lasts six to eight weeks with one or two harvests per week. It is customary to have two crops in the forcing shed to complete the forcing season. Once used, the roots are discarded in order to prevent serious spread of diseases. A new batch of rhubarb plants should be available each year for forcing.

In outdoor production, rhubarb should not be harvested the first year and only for a short time during the second year to permit the roots to build up food reserves. Beginning with the third year, a harvest period of eight to ten weeks can be established. Seed stalks should be pulled whenever they appear.

Rhubarb can also be grown in greenhouses under a full-light forcing regime. In all cases only the most vigorous plants are selected for continued growth.

Pest Control

Local recommendations for pest control should be sought before planning to use chemicals to combat rhubarb pests.

Weeds: The control of grass weeds in rhubarb can be accomplished with glyphosate or Round-up or with Grammaxone or Paraquat.

Insects: The rhubarb cucurlio is a serious pest. Destruction of wild dock plants in the area of rhubarb plantings helps in its control. Other insects are potato stem borer and nematodes.

Diseases: Crown rot and virus diseases are ever present diseases of rhubarb. The removal of diseased plants from any planting is a major control method. For the forcing rhubarb operation, the prevention or control of botrytis leaf rot is essential. Removal of plant residues from previous crops in forcing sheds is essential. It is imperative to consult with local specialists about recommendations for using registered chemicals.

Harvesting and Storage

Field Rhubarb. The leaf blade is detached just at the petiole juncture with the leaf-only the stalks are marketed. Good rhubarb storage is linked to preventing the loss of moisture. At high RH and O°C to 2°C, storage life may be four weeks; two weeks at 4°C, and one week at 10ºC.

Forced Rhubarb. Forced rhubarb is a very delicate, perishable commodity requiring great care in handling. Any rough handling detracts from the appearance of rhubarb due to blemishes that will occur. The temperature guidelines for the shelf life of field rhubarb apply to the forced crop.

CHAPTER 20

Rutabaga (*Brassica napus* var. *napobrassica.*)

Family: Cruciferae

Chromosome number: 2n= 38

Origin

Gerard, the father of English herbalists, during the time of Henry VIII, gave the first reference to the turnip or rutabaga. Gerard stated, "It growth in fields and diverse vineyards, or hoppe-gardens in most places in England. The bulbus or knobbed rotte, which is properly called rapum or turnip, and hath given the name to the plant, is many times eaten raw by the poor people, but most commonly boiled." According to Sturtevant, the rutabaga was introduced to North America at the beginning of the nineteenth century. It was then called Swedish turnip, Lapland turnip, or rutabaga. There are two forms, the yellow fleshed and the white fleshed. It was referred to in England as turnip rooted cabbage. The history of the yellow-fleshed type in North America is not clear but the so-called Swedish turnip or rutabage, probably yellow-fleshed, was mentioned by McMahon as growing in American gardens in 1806. The rutabaga was known in Europe long before it was grown in England. It had spread to France and southern Europe by the seventeenth century.

Botany

Rutabaga or Swedish turnip closely resembles the common turnip B. rapa. Rutabagas are now predominantly yellow fleshed. Although the hypocotyl root system resembles that of the turnip, rutabagas have a short stem or neck at the upper part of the hypocotyl, a distinguishing feature that separates it from the turnip. The flesh is solid and its color remains after cooking. The early leaves are bluish white with thick fleshy petioles. The leaves on rutabagas are smooth and free from hairs. The yellow flowers are relatively large as compared to turnips. Although like other members of the Brassica family subspecies the rutabaga readily cross pollinated crop, the flowers are self-compatible. Any roots left to over winter that subsequently goes to seed will readily pollinate a supposedly isolated seed steckling field within one half mile or so. Left unhampered by man's control, the rutabaga in a few short generations may revert to producing a non-fleshy root. Indications clearly are that close control of pollination sources must be adhered to with this vegetable when quality seed is required.

Consumption and Uses

The rutabaga is a readily adapted, economical, tasty vegetable available throughout most of the year. The easiest way to prepare a rutabaga root for cooking is to slice off the top with a good sharp kitchen knife, lay the flat surface on the cutting board, and cut as required for recipes. The flavor of the rutabaga makes it suitable not only as a cooked vegetable but also in soups, salads, and baked goods. Fresh uncooked rutabaga can be shredded into salads.

Nutritional Values

Nutritional values of the rutabaga are presented in as follows:

Table: Mineral and Vitamin Content of the Rutabaga (100-g edible portion)

	Ca (mg)	P (mg)	Fe (mg)	Na (mg)	K (mg)	Vito A (I.U.)	Thia-mine (mg)	Ribo-flavin (mg)	Nia-cin (mg)	Ascorbic Acid (mg)
Raw	66.0	39.0	0.4	5.0	239.0	580.0	0.07	0.07	1.1	43.0
Cooked (boiled), drained	59.0	31.0	0.3	5.0	167.0	550.0	0.06	0.06	0.8	26.0

Source: United Fresh Fruit and Vegetable Association 1970.

Major Cultivars

There is no cultivar of rutabaga in India. The main cultivar in the United States is American Purple Top, a rank, rapid grower. In Canada the main cultivar is Laurentian. Because Laurentian carries no genes for club root or mosaic resistance, plant improvement has aimed at transference of resistance to these diseases from closely related species. Several club root resistant cultivars have been bred, York and Fortune being the two most commonly used. They are phenotypically Laurentian types. The ideal Laurentian weighs about 700 g and has a circumference of about 10-12 cm.

Recent breeding emphasis in the Laurentian cultivar has focused on incorporating triazine resistance, so that rutabagas can be grown immediately following corn in the rotation or used as a clean-up crop regardless of previous weeds or herbicide applications. The triazine resistant gene of the canola has been imparted to the rutabaga. The future direction of rutabaga improvement lies in combining all the major disease resistances, including mosaic and club root, with triazine resistance. Should the need arise, a further option will be to develop hybrid vigor so as to stabilize the integrity of the previous breeding efforts in a single hybrid.

Less important cultivars are Golden Table, Victory, Neckless, and Ditmar.

Climatic Requirements

The rutabaga is a cool season vegetable. It should be emphasized that when mean temperatures remain high (above 25°C) for prolonged periods during the rutabaga's growth cycle, the root grows too quickly, becoming unshapely and subject to severe growth cracks. Should soil moisture also remain near capacity, the volume of unacceptable oversized roots can become excessively large. If rutabagas grow too large (in excess of 12-14 cm in diameter) and exhibit growth cracks and roughness, they become unacceptable for the fresh market. They are then used for soups, chips, or livestock fodder. As a consequence of this high-temperature response, rutabagas are sown in areas that rarely become too warm for prolonged periods. As the mean temperature rises, the seeding date is coordinated with the optimum harvest date. Rutabaga flavor peaks after the plant has been exposed to early fall frosts.

Soil Requirements

Rutabagas can be grown on wide range of soil type, if the proper nutrition is provided. A medium loam soil probably imparts smaller swings in nutrient balance than either sandy or clay loam soils. In addition, a medium loam soil holds moisture in better balance. Thus the most appropriate soils are those that are not extreme in classification or nutrient balance. Since rutabagas rarely need to be sown early in the growing season, the requirement for early-warming spring soils, such as the sandy types, is not imperative for this crop. Sandy soils are in fact the most difficult from which to extract a maximum uniform crop of rutabagas.

Fertilizer Requirements

Rutabaga nutritional requirements are best worked out with a reliable soil test obtained prior to field application of fertilizers. The nitrogen requirement for rutabagas is about 35 kg/ha. On soils with Mg of less than 100 kg/ha and/or with pH values less than 6.5, the amount of K can be cut in half.

In general rutabagas fitted into a rotational pattern do not require large amounts of fertilizer, particularly if the previous crops were well fertilized. Farmyard-manured soil should not be applied prior to or during preparation for rutabagas. Too much organic manure and/or excessive N causes poorly shaped rutabaga roots to develop. Rutabaga nutrition must be carefully programmed to anticipate the growth needs of the crop and prevent excess growth leading to cracking.

Minor Elements: Rutabagas may develop water core or brown heart, a physiological breakdown that appears initially as brownish water soaked patches, which increase in size until most of the root is affected. Some severely affected rutabagas disintegrate internally, a condition undetected until they are in storage. This condition usually occurs in soils with a low soluble Bo content and a pH greater than 6.5. Applications of boron broadcast directly to the soil do not always overcome the deficiency. A spray consisting of boron solubilized in water and applied three or more times helps to prevent this deficiency. Where Bo deficiency occurs, local or regional recommendations should be sought.

Cultural Practices

Except in the coolest short-duration production areas, rutabagas for winter storage are seeded around mid-June. This allows ample time to prepare a sound, weed-free, moisture-holding seedbed. There is no advantage to overworking the soil with excessive deep plowing and/or cultivating, as this robs the seedbed of essential moisture. A firm, moist seedbed sets the stage for singulation precision seeding, at a depth of 3.0 to 4.0 mm and spaced every 15 cm. If uncertainty of moisture or seed quality exists, the precision seeder can be set to drop a single seed every 7.5 cm, and if necessary every other seedling can be blocked out later in a single thinning pass through the field. Distance between rows will vary from 60 to 90 cm depending upon equipment and wheel spacing.

Rutabaga seed should be closely sized to promote uniformity in the seed rows. Since large seeds always germinate faster than smaller ones and reach full maturity size as much as five to six weeks earlier, they should never be mixed. To provide a high degree of uniformity, it is best to hand screen the seed. While maintaining uniformity within each row, it is possible to alternate rows of different seed size in the field for successive harvests.

Irrigation and Drainage

Throughout the main rutabaga growing areas, natural rainfall patterns are usually sufficient to carry a crop without serious water stress. In loamy soils rutabagas are generally able to survive between summer rains. Only in sandy soil is supplemental irrigation normally required. Minor element deficiencies may surface if too much irrigation is applied to rutabagas. Rutabagas do not perform well if the soil is not well drained throughout the root area.

Weed control: Rutabaga production is threatened by grasses such as quack grass, barnyard grass, and green foxtail, although the normal late spring ploughing or land preparation slows down the inroads of weeds. Another damaging weed is the crucifer shepherd's purse. Because rutabagas are grown in rows, clean cultivation can be maintained even though with each successive pass of the cultivators a new crop of weed seeds is brought to the surface. Herbicides are used if necessary, based upon regional recommended practices. Certain herbicides can damage rutabagas. Never use a sprayer that has previously been used to apply 2, 4-D-derived compounds, MCP, or related products. Except for certain recently developed cultivars, rutabagas are highly sensitive to triazine or trifluralin derivatives and show greater than usual sensitivity to herbicide spray drifts from neighbouring agronomic fields. Progress is being made in developing spray material and/or injury resistance, as has been explained.

Insects: The rutabaga is highly susceptible to the cabbage maggot. The adult, who resembles a small house fly, lays its eggs in the soil near the young plants. These hatch into small white maggots that feed on rutabaga roots, rendering them unmarketable. Once again local or regional insecticide recommendations should be followed. In some areas the use of non drift, subsurface granular applications of insecticides is recommended, whereas in other areas sprays or drench treatments are preferred. Another destructive insect pest is the flea beetle, which attacks the emerging rutabaga seedling both below and above ground.

Diseases: Rutabagas are highly susceptible to the transference of viruses, principally the turnip mosaic virus (TMV) by the green peach aphid. Constant crop monitoring to detect and destroy aphids and other insects is a mandatory operational practice with rutabagas.

Mosaic-loaded roots do not store well due to breakdown by rot organisms. Next to the TMV the most serious disease attacking rutabagas is club root. Careful rotations with rigid standards of field sanitation will reduce infestations or hold club root in check. Five-year rotations are minimal for rutabagas. Resistance to club root has been developed, but since strains of club root differ from region to region, there is no assurance that resistance to anyone strain is universal. The disease builds up in an area over time and can become a seriously limiting factor for rutabaga production, particularly in the cooler regions.

Seed-borne rutabaga diseases are dealt with by means of hot water treatments and/or seed dressings capable of protecting the germinating seedling. These diseases include black rot, neck rot, rhizoctonia, scab, and black leg. Sound, carefully planned farm practices prevent many serious pest problems from getting out of control.

Harvesting and Storage

With the advent of precision seeders and mechanical harvesters, almost all rutabaga acreage is mechanically harvested. The harvesters top and dig the rutabagas, shake loose excessive soil, and feed the roots directly into hopper-like dump trucks to be taken to storages. A good practice is to submit the roots to short inspections on a conveyor belt prior to consigning them to final storage. Throughout the harvest process, care must be exercised in operating the harvesters and unloaders to minimize root damage, as bruising reduces the storage life of rutabagas.

Rutabagas may be piled 2 to 4m deep on slatted floors with forced air circulation. In order to reduce moisture loss, perforated plastic sheeting is sometimes placed horizontally at 45cm intervals. For best storage results the temperature should be held close to 0^0C with relative humidity maintained no lower than 90 percent. Well-stored rutabagas can be kept in prime condition for upwards of six months without requiring sprout retardants.

Prior to shipping, rutabagas are trimmed, washed, graded, and waxed. Special driers are used prior to the final waxing. Rutabagas maintain a longer supermarket shelf life if hot-wax treated. They shrivel less and if the roots are blemish free, they will maintain their appearance for several weeks longer.

CHAPTER 21

Spine gourd (*Momordica dioica* Roxb.)

Family: Cucurbitaceae

Chromosome Number: 2n=

Botany

Momordica dioica Roxb. is a perennial with tuberous roots, stem is sylendrical, branched, furrowed, glabrous and shining. Leaves are membranous, broadly ovate in outline, variable, cordate at base more or less deeply 3- 5 lobed, distinctly denticulate. fruit long, ellipsoid, shortly beaked, densely echinate with soft spine.

Nutritive Value

The delicious young tender fruits are used as vegetable. Spine gourd fruit, per 100 g edible part, contains moisture 84.1g, fat 1.0 g, fibre 3.0 g, carbohydrates 7.7 g, carotene 162 mg, protein 3.1 g, calcium 33 mg, phosphorus 42 mg, iron 4-6 mg, thiamin 0.05 mg, riboflavin 0.18 mg and niacin 0.06 mg (Ram et al.,2004).

Climate and Soil

Spine gourd is a warm and low humid season crop. It can be grow both in tropical and sub tropical regions. This crop requires better sun shine for its growth. This can be grown successfully in areas where temperature ranges 25-40^{0}C with annual rainfall of 150-200 cm. It grows well under sandy loam to clay soil soils, rich in organic matter. Optimum soil pH for the crop ranges from 6.0-7.0.

Field Preparation

The field is prepared by digging pit. The pit is filled with 5kg FYM, 150g SSP and 50g MOP and 3g Furadon to prevent termites. Two time topdressing of Urea @ 80g near the root zone promote plant growth and development.

Seed Rate: 3000-5000 tubers/ha.

Time of Sowing: The favorable time for planting is during February-March.

Spacing: 2mX 1 m.

Manure and Fertilizer: The dose of fertilizer depends upon soil, climate and system of cultivation. Well rotten FYM @15-20 t/ha is applied at the time of land preparation. NPK @ 120:80:80 kg/ha is added with full dose of P and K before sowing and half dose of N at the time of vining and remaining half dose of N before flowering.

Irrigation: Just after sowing of seed or placement of cuttings irrigation is essential and it should continue for 4-5days.

Training: As the crop is climbing in nature, provision of supporting structure encourage proper vegetative growth and thereby contribute to good yield of quality fruits. Different system of training viz. single stake system, bower and kniffin system are followed as per availability of training materials.

Harvesting and Yield: Fruits are ready for harvest within 75-80 days in first year and 35-40 days from second year. Fruits should be harvested at mature green stage. Under optimum crop management condition fruit yield is 75-100 q/ha.

CHAPTER 22

Sweet corn (*Zea mays* L. var. rugosa)

Family: Gramineae

Chromosome number: 2n= 20

Sweet com, *Zea mays* L. var. *rugosa*, earlier known as *Zea mays* L. var. *saccharata*, belongs to the monocotyledonous group of plants-Gramineae (grass) family. It is also known as papoon com and maize.

Origin

Sweet com is a plant of the New World characterized in the immature state by its sweet kernels. Evidence points to mankind's familiarity with com over thousands of years. Findings in excavated Mexican tombs in Techuacan Valley point to wild com cob remnants dating back to 3400-5200 B.C. and cultivated corncobs from 3400-2300 B.C. Fossilized com pollen grains estimated to be 80,000 years old have been found in Mexico as well.

Peru was probably the primary source of domesticated com, while wild com was thought to have originated in the lowlands of what is now Central America and northern South America, an area of rain forests and savannahs characterized by intense rainy seasons and long dry periods. Likely hybridization occurred within the types of wild grass like corns and maize. The first types probably were like present day pod-corn-each individual kernel was covered by a husk that pro-vided a survival mechanism not now found in com. Modem com, with its tight protective husk firmly attached to the cob, has no means of dispersal and would probably not last more than a generation or two in the wild. Self-seeded com found to have over wintered develops a clump of plants quite incapable of producing viable individual kernels for very many generations.

In the regions where com was first grown, sweet com types must have surfaced from time to time. Apparently these chance-occurring sweeter types were not to the liking of the people who used com as a main diet, since very little attention was given to their development (or they may have been harder to perpetuate). Similarly, the people in the southern United States who early accepted com in their diet gave little thought to the sweet or sugar corn, probably because it was different in taste.

Sweet corn has been grown in the northern United States and Canada since 1779. In that year a Lieutenant Richard Bagnal, who accompanied General Sullivan on his expedition against the Six Nations Indians, brought back a new kind of corn from the Susquehanna region of Pennsylvania. Called "papoon" by the Indians, it had white, wrinkled, glassy seeds and a reddish cob. This sweet corn, which was quickly accepted in the northeastern region, led to the first offering of a commercial sweet corn cultivar-the Sugar or Sweet-in 1832. The rapid introduction of sweet corn cultivars beginning in 1850 became the foundation for the modern sweet corn breeding programs. Since 1832 over 2000 cultivars have been introduced, moving from predominantly white varieties until the introduction of Golden Bantam in 1902, a cultivar that brought about a gradual swing to yellow sweet corn. With the introduction in the 1970s of the high sugar and super white types, white kernel corn is again being popularized. A chronology of the most important early cultivars and their modern derivatives is given in Table 14-1. The major changes in sweet corn have been the increases in ear size, depth of individual kernels, number of rows, and overall yield, together with the development of super sweet corn.

Botany

Sweet corn is an annual, herbaceous grass type plant distinguished from other corns by its single gene for high sugar content. When the sweet corn seed is dry, it is wrinkled and translucent (glassy) and has few or no starch grains.

Consumption and Use

Sweet corn is a popular vegetable. The U.S. 1978 per capita consumption of sweet corn was 3.40 kg fresh corn, 2.72 kg canned, and 0.82 kg frozen, for a total of 6.94 kg.

Probably corn on the cob (boiled, steamed, or roasted) is the favorite use of sweet corn. It is also canned whole kernel, vacuum packed, creamed (cream style), and frozen (whole kernel and on the cob). Sweet corn is part of numerous dishes ranging from specialty items to pickled products.

Reproduction: The plant is monoecious; that is, the staminate inflorescence (male flower parts-the tassels) is separately located at the terminal of each main shoot, whereas the pistillate inflorescence (female parts-the ear) is made up of the cob, rows of eight or more kernels each having its own silks (styles) and covering husk (bracts) located laterally on the main stalk always well below the tassels. Sweet corn is normally cross-pollinated-only 1 percent is self-pollinated by wind or gravity. The tassels always mature and shed pollen several days before the stigmas (silks) become receptive. Because the stray pollen of other types of corn can materially alter the uniformity (color) or increase the number of undesirable starchy kernels, it is important not to grow sweet corn near the other corn types. Both culinary and visual qualities of sweet corn can be seriously impaired if adequate separation is not provided when planting divergent types of corn in the same general area.

The term *xenia* refers to the immediate influence of male pollen on the seed of another strain. For example, variable kernel color, carried via male pollen, may be evident in the seed as the result of accidental pollination. Usually this phenomenon occurs when two cultivars such as sweet corn and field corn or sweet corn and popcorn are planted side by side or in adjacent fields. The following can happen: non starchy (sweet corn) crossed with starchy (field corn) always ends up starchy; non yellow crossed with yellow shows yellow; non colored aleurone (protein) layer crossed with purple gives purple; and non colored aleurone layer crossed with red gives red. *Xenia* is considered an undesirable side effect in sweet corn, unless of course a mixture of yellow and white is sought.

Roots: The corn plant roots are initiated from the root crown, which consists of whorls numbering from two to ten, one above the other, with very short internodes. The roots are either main verticals or main laterals. The former grow briefly to the side but then quickly move downward whereas the latter roots emanate from the crown horizontally for a distance and then begin a downward course. The net result is a massive spreading and deeply penetrating fibrous root system occupying the upper two meters of soil under each plant. In addition to the roots described, at about the time of tasseling the corn plant also produces aerial roots known as prop or brace roots initiating from the whorl above the soil line. Above the soil these roots are thick, heavy, and outward growing; when they penetrate the soil, they grow downward, forming a profusion of fine roots that serve as anchorage to the plant as well as absorption organs similar to the main root system.

Stem: Great variation exists in the height of mature sweet corn plants, which range from dwarf-60 cm high-to very tall-300 cm or more in height. The stems are jointed as in all grasses. The internodes are filled with vascular bundles (pith) or fibers, and the nodes themselves are solid. For the most part main line hybrids do not sucker as did the majority of earlier sweet corn cultivars.

Leaves: Like most grasses, sweet corn has leaves that are alternately arranged on opposite sides of the stem. They vary in number from eight to twenty. The leaf is composed of three parts: The first is a cylindrical sheath split down one side with edges more or less overlapping throughout its length. The second is a long, flat, strap-shaped blade with a strong midrib and numerous smaller veins parallel to it. The third part is a thin membranous collar, the ligule, which arises at the upper side of the top of the sheath and fits tightly about the stem. The leaves are structured so as to protect the plant from excessive transpiration by rolling inward and exposing a smaller leaf area-' thus reducing the evaporating surface.

Hybrid Sweet Corn

Until 1933, when G. M. Smith introduced Golden Cross Bantam, there were no true hybrids of sweet com. There were many controlled crosses of cultivars already in existence (see Table 14-1), however. Smith set about developing two genotypically different inbred lines of Golden Bantam identified as P39, the seed parent, and P51, the pollen parent. The resulting progeny, which were high-quality, productive, bacterial wilt resistant, and highly uniform in maturity, very quickly established the direction sweet com breeding would take. Through this highly successful breeding program, terms like "heterosis," "hybrid," and "hybrid vigor" became familiar. Many combinations of breeding techniques were tried. What remains today is the significance of the single cross method and effective inbreeding to stabilize traits of sweet com. When the inbreeding lines were well chosen and stabilized, the hybrids that came from using the higher-yielding inbred lines as the female parent produced the best progeny. The highest tribute to the success of this first sweet com hybrid development is that today one can only with difficulty distinguish seed catalogue listings of open pollinated and single-crossed sweet com lines.

Major Cultivars

Many factors influenced the evolution of present cultivars aside from the desire for high yield, uniformity of ear size, and better quality. Cultivars available today have multipurpose uses. For instance, processors demand multiple rows of deep kernels to increase the ratio of kernel weight to cob weight, thinness of the pericarp to impart greater tenderness, input of the high sugar gene to prolong the sweet taste of com whether it remains on the plant a day or two past normal maturity or is removed and stored until ready for use.

In addition, breeders have incorporated genes conferring resistance to such diseases as southern wilt. By transferring to sweet com the high ear attachment of field com, breeders gave an impetus to mechanical harvesting, which is now the only way to harvest sweet com whether for fresh or processor use. Since the South rarely grew sweet corn because of the rapid conversion of sugar to starch under high temperature growing conditions, sweet corn production never moved to that region until the super-high or high-sugar lysine gene transfers were accomplished. The starch gene was replaced with the sweet or sugar gene, essentially eliminating the formation of the starch and firmly establishing sweet corn production capability from north to south. Virtually every day of the year in the United States and Canada it is now possible to offer fresh corn on the cob that is attractive to look at and good to eat. In addition, high-sugar cultivars are now available in yellow, white, or bicolor (white and yellow).

Traditionally sweet corn processors made their own arrangements for contract crops, and so cultivar listings are generally not available. Nevertheless new and not so new sweet corn hybrids for roadside corn on the cob and the home garden are available. To be sure of getting regionally adapted cultivars, one should check with local, state, or provincial extension personnel for the latest recommended list of sweet corn cultivars. Usually the recommended listings will separate these into shipping and roadside and home garden use.

This section would be incomplete if mention were not made of the super sweet cultivars. These are listed as Illin Extra Sweet, Early Extra Sweet, Butter Fruit, Florida Stay Sweet, How Sweet It Is, Kandy Korn EH, and Sweet Temptation. In the bicolor group can be found Bi-Queen, Honey'n Frost, Sprite, Burgandy Delight, Harmony, Sweet Sue, Sweet Sal, and Early Gold. In addition, there are new and established hybrids with very distinct temperature adaptations.

Geographic/Climatic Requirements

Sweet corn is a plant with a long history of human manipulation, and never is one more aware of that than when looking at the plant's wide adaptation to regional environmental conditions. Sweet corn can be found from 58°N to 400N latitude and from sea level to 4000 m altitude. Although one should not seed sweet corn in soil temperature lower than 10°C, the first vegetative shoots are capable of withstanding a late spring frost, provided the heart of the shoot remains undamaged. Sweet corn requires bright hot days to yield well. Now that the high sugar gene is established, high temperature and humidity present no drawback for fresh market sweet corn production in any geographic region-optimum temperatures range from 15° to 35°C. Similarly, neither long days of twenty-three hours in the far north nor the short twelve-hour days of the tropics have detrimental effects on this plant.

Soil and Fertilizer Requirements

Sweet corn requires a medium-heavy, well-drained loamy soil well supplied with organic matter. The lighter sandy soils should be avoided unless irrigation is readily available with a good fertilizer program. Sweet corn fits well into a four-year rotation that includes a legume or green manuring crop. If livestock fits into the farm operation, well-rotted manure at about 40 tons per ha would be advantageous. Soil tests should be taken of any field prior to working out the fertilizer requirements of sweet corn. Advice should be obtained locally as to the best formula or rate of fertilization. Sweet corn is a heavy feeder. Ten tons of stover and five tons of ears will take up 52 kg of N, 15 kg of P, and 28 kg of K per ha. Response to a good soil situation with adequate nutrients usually produces high yields of sweet corn. Effective fertilizer utilization is possible by banding the fertilizer below and to the side of the seed row. As a general rule sweet corn response to the addition of micro nutrients to the soil is low for boron and molybdenum, medium for manganese, copper, and iron, and high for additions of zinc.

Cultural Practices

For maximum yields of any vegetable crop, great care is required in the preparation of the seedbed to insure the best possible uniformity both with regard to the soil moisture and the flow of the seed equipment throughout the seeding process. The seeding equipment should be set to the correct size and plate for sweet corn as reckoned from the seed bag tag, which identifies seed size and the appropriate plate opening. Earlier writings stressed the removal of unwanted suckers in sweet corn, but for the most part modern sweet corn hybrids do not sucker.

For processing corn, the grower should be guided by the instructions provided by the company field staff to time the maturity of corn with the daily capacity of the processing plant. As a rule heat units (degrees above a daily mean of 10°C), plus the average daily temperature for the week in which the cultivar is expected to reach the milk stage, will dictate how much of a particular cultivar is sown in the spring and what cultural practices will be required to meet the target for optimum plant production. The processor seeks to optimize cut kernel ratio to total ear yield according to the use of the processed corn-whether it is to be frozen, canned, or processed cream style. As a rule the best quality corn is frozen or used whole kernel vacuum packed, and the late milk to early dough stage goes into the cream style corn. Spacing for processing corn is usually closer in and between rows than is required for fresh corn production.

For market or fresh corn, which may also be grown under contract with shippers, the requirements for harvesting quality corn are more stringent and more closely controlled. Cultural practices are geared to move the fresh corn as early as possible to the market. Fresh corn may be transplanted-using various techniques to produce undisturbed transplants-or field planted when the worst dangers of frost are over. Other techniques involve use of both clear and black plastic mulches or seeding in deep furrows and then covering with clear plastic to avoid frost damage. Another method when sowing in the chilly late winter on the California deserts is to ridge the fields running north and south, seed the corn on the south-facing side, and top the ridge with some form of temporary windbreak. The use of cloches or plastic tunnels will also gain some time advantage for the early market by conserving heat.

Whether sweet corn is grown for processing or fresh market, cultivars can be selected to come to maturity in an orderly sequence to match the market requirements. A single cultivar can be seeded sequentially to achieve the same purpose. This latter method may be more risky than picking the right seasonal cultivars, which as a rule will adapt to hot or cool spells better than one cultivar with a range of seeding dates.

Irrigation and Drainage

Sweet corn is very sensitive to the water supply in the soil. No matter where this crop is grown, conditions arise during the growing cycle when the water supply drops to dangerously low levels. In the desert or semiarid regions, it is enough to provide a weekly or ten-day irrigation water replacement regime depending upon the soil type. In the more humid eastern region, where rainfall can be irregular, droughty periods often occur. Since sweet corn as a rule is grown in a mixed cropping operation where irrigation is in place to supplement rainfall, it is justifiable to irrigate sweet corn. (Sweet corn by itself cannot justify an expensive supplemental irrigation system.) Some dry years have left sweet corn badly damaged, and in some cases plantings have been completely destroyed by drought in the mid-west and eastern parts of the continent. In Florida where sweet corn is a fall, winter, and spring crop, irrigation is essential.

Most irrigation of sweet corn and other vegetables is done by highly mobile pivot systems that provide fixed circular coverage within the radius of the sprinklers. Trickle irrigation systems deliver restricted water to each plant, and in the far west, pumping systems deliver water to a border dyke system that allows water to flow by gravity through the fields so that the beds become water laden.

The use of overhead sprinklers must be carefully managed so that during pollen release and silking the driving force of the water droplets do not interfere with the pollination process. Under the cool, dry conditions of the northern plains and western Canada, where irrigation is essential for a good corn yield, temperature conditions are low enough to cause super cooling of the micro environment of the sweet corn plant and prevent proper pollination and ear production. Under a regularly scheduled irrigation program set up to supplement the rainfall, the super cooling effect may in fact become deleterious to the formation of well-filled ears of corn. Since sweet corn is a wind-pollinated crop this type of phenomenon must be taken into account when planning an irrigation program. Obviously the same applies to the effects too much or too frequent rainfall can have on the maturation processes of sweet corn.

Water levels of soils cannot go to extremes without stressing the sweet corn crop. As with all vegetable crops, sweet corn cannot tolerate excessive water held in the root zone. Sandy soils, especially under irrigation regimes, can quickly develop impervious zones preventing excess water from percolating away from the root zone. Similarly, heavier soils often do not give up water fast enough, resulting in slower growth. Tile drains are essential in the humid eastern parts of the continent, where this problem is common.

Pest Control

The specific recommendations for the use of chemicals and cultural practices to control pests must be strictly adhered to especially for a crop like sweet corn, which is so universally grown. Experiment stations and extension advisors should be consulted before planning a pest control program for sweet corn.

Weeds: Few vegetable crops are as dependent upon chemicals to control weeds as is sweet corn. The same guidelines for field corn evaluation in states and provincial annual herbicide recommendations apply to sweet corn. Some precautionary guidelines with respect to sweet corn weed control are concerned with species that will have to occupy the land subsequent to sweet corn production. For example, the use of atrazine compounds is very popular with corn growers, but unfortunately most broad-leaved crops are highly sensitive to very small amounts of residue. Therefore smaller quantities are applied to reduce the risk of soil residue holdover, or else other less efficient herbicides must be introduced. Failing the all out chemical route, it is recommended that weed control practices such as cultivation be used wherever possible. Because sweet corn is a row crop, good weed control can be achieved without seeking the absolute chemical control so highly regarded in corn-growing circles. Since sweet corn is now fully mechanically harvested, weeds cannot be permitted to become large enough to interfere with mechanical harvesting efficiency. If chemicals are used to control weeds in sweet corn, they must be effective against a wide array of grasses as well as broadleaf weeds both annual and perennial.

Insects: A host of insects seek out the sweet corn plant from the time the seed goes into soil until the crop is harvested. In particular, seed maggot, wireworm, flea beetle, European corn borer, ear worm, aphids, fall armyworm, and corn root worm infest the sweet corn and can effectively damage the crop. Local recommendations to control insects should be followed explicitly.

Diseases: Sweet corn diseases include Stewart's bacterial wilt, which attacks the young plant; leaf spot including northern corn leaf blight; common rust in the north, especially under cool, humid conditions; southern corn rust; smut, an increasingly serious stem and ear disease; sorghum downy mildew, a serious southern problem; crazy top (a form of downy mildew); maize dwarf mosaic; corn stunt; and wheat streak mosaic. Some tolerance or resistance exists to some of the various diseases listed in specific cultivars-for example, in Apache and Aztec.

Aztec is tolerant to bacterial wilt, common smut, and southern leaf blight. Apache is tolerant to bacterial wilt, common smut, northern leaf blight, southern leaf blight, and maize dwarf mosaic virus. Similarly there are many cultivars resistant to Stewart's wilt.

Harvesting and Storage

Whether for processing or fresh use, sweet corn must be harvested at its peak quality. Approximately twenty-one days after the silks have appeared, the ears are ready for harvest, given no unusual weather conditions such as drought or coolish wet weather, both of which may change the rate of maturation.

Processors use various devices to determine the prime maturity of sweet corn. The most commonly used are the following:

- ***Refractometer-*** *an* instrument used to measure the total soluble solids as an index of maturity (a small portable one carried by the field staff or a more sophisticated version)
- ***Succulometer- a*** laboratory device that takes field samples and measures the amount of juice (a few percentage points above or below a reading of 70 will indicate maturity of the kernels)
- ***Steinlite, Brown-****Duvals, Brabender oven-* an instrument that provides an objective moisture content measurement
- ***Lee-Kramer-Shear Press-*** *a* laboratory instrument that measures the firmness or tenderness of the kernels (much as the texture meter does with peas), giving an objective measurement of tenderness of corn kernels

A great deal hinges on accurate quality measurements taken up to the time of harvest to insure that the best possible quality sweet corn reaches the plant.

Shipping sweet corn must also be of peak quality. Growers and experienced field personnel can fairly accurately measure the maturity of sweet corn by the look of the maturing ear, the feel of the tightness of the husk to the tip of the ear, and the condition of the silk. A hand refractometer can be used as a guide as well, but it damages the ears if physical tests are made.

The harvesting of sweet corn used by both processors and fresh packers is done by one- or two-row mechanical harvesters. The most advanced machines cut rather than snap the ears to avoid bruising the kernels. The newest harvesters are one-man operations equipped to harvest at night, when the field heat of the corn is at its lowest level. All mechanical harvesters are provided with a hookup for trailing tractors or have on-board hoppers that elevate the ears and empty them into trucks or trailers alongside the machine. Rarely is sweet corn hand harvested in the field. The present machines do a better job of harvesting sweet corn than was done by the hand harvest operations. Harvested sweet corn should not linger in the field, because the rate of respiration (particularly of the husk and cob) is rapid at 21°C and twenty times as rapid at 33°C.

Although sweet corn is seldom stored for long, the temperature of the harvested corn should be reduced to 0°C as quickly as possible. This is usually done by hydro-cooling the corn by passing it in crates or bulk through a slow-moving tunnel conveyor system, which sprinkles iced water continuously on the corn until it emerges from the tunnel. It requires about twenty-five minutes to lower the temperature of an ear of corn from 300e to 4.4°C. Sweet corn with a field temperature of 21.1°C generates 64,000 Btu per ton per twenty-four hours; when cooled to 0°C the rate of respiration is reduced to 9,000 Btu. However, even when cooled to 0°C the heat of respiration persists.

Sweet corn properly cooled and held at around 0°C has a quality shelf life of about six to eight days. If ears of corn are partially or wholly displayed unhusked, some form of perforated film must be placed around them not only to prevent moisture loss but also to dissipate the heat of respiration and gases, and they must be held at refrigerated temperatures. Regrettably most supermarket produce managers do not fully understand the critical physiological changes occurring in sweet corn and hence all to often market deteriorating corn. The public is easily disenchanted with a corn product that fails to offer a quality treat.

CHAPTER 23

Swiss chard (*Beta vulgaris L. var. cicla*)

Family: Chenopodiaceae

Chromosome number: 2n= 18

Swiss chard, or *Beta vulgaris* L. var. *cicla mog*., is a member of the Chenopodiaceae (goosefoot or pigweed) family. It is also known as leaf beet, seakale beet, silver beet, spinach beet, or simply chard.

Origin

Chard is a very old vegetable that originated in the Mediterranean basin. The ancient Greeks (350 B.C.) recognized the red chard, the white (call sicula), and the black (so named for its intense deep green color) the last was the most highly regarded. Theophrastus, Aristotle, Descorides, Eudemus, and Athenaeus all had something to say about this vegetable. Similarly the Roman Pliny made frequent mention of chard, as did the herbalists of Europe later on. The wild forms grow on the Canary Islands. The root (red beet) of the chard was first used medicinally before its culinary values were known. Chard was introduced to North America around 1806 by European immigrants. It is a unique species valued both for its cooked greens and its cooked petioles, but is primarily grown in home gardens rather than commercially in North America.

Botany

Like the beet, chard is a herbaceous biennial, botanically very similar to the table beet except that it does not form an enlarged hypocotyl. The plant develops a compressed cluster of stalks similar to celery, forming distinctive enlarged petioles with heavily savoyed leaves. The distinctive color of chard is due to the petioles, which may be white, red, or pink. Swiss chard is characterized by dark green leaves. The flowering habit and plant characteristics are similar to the table beet. Unlike the red beet, however, chard does not respond to high temperatures, which cause bolting in beets, spinach, and lettuce, in the first year as well as in subsequent years.

In order to induce flower stalk formation, chard roots must be exposed to extended periods of chilling. Because chard is a highly vegetative plant, as the first-formed or outer ring of petioles are harvested, new heart growth occurs and the outer petioles successively mature for continuous harvest. If the outer petioles are not harvested promptly, the expansion of new ones is inhibited.

Consumption and Uses

Chard is a very minor crop grown in relatively small blocks no more than a third to one-half hectare in size due to the highly intensive labor requirement. Chard is used as a cooked product-the stems are used like cooked celery and for best flavor should be steamed. The leaves are cooked the same as spinach (with only the water on the leaves left from washing them). Although spinach and chard have distinctive strong flavors, when the two are combined and cooked together the flavors offset each other and produce a pleasant blend.

Table 17.4. Mineral and Vitamin Content of Swiss Chard (100-g edible portion)

	Ca (mg)	P (mg)	Fe (mg)	Na (mg)	K (mg)	Vito A (I.U.)	Thia-mine (mg)	Ribo-flavin (mg)	Nia-cin (mg)	Ascorbic Acid (mg)
Raw	88	39	3.2	147	550	6,500	0.06	0.17	0.5	32
Cooked (boiled), drained	73	24	1.8	86	321	5,400	0.04	0.11	0.4	16

Source: Watt et al. 1963.

Major Cultivars

Lucullus, a yellowish-white-petioled chard, is the most important cultivar. There are many others including highly colored ones, such as Fordhook Giant, Burpee's Rhubarb Chard, Giant Lucullus, Rhubarb Chard or Ruby Red, and Dark Green White Bibbed.

Geographic/Climatic Requirements

Swiss chard grows throughout all gardening areas of the United States and Canada. Having the ability to resist high temperatures by not bolting is a distinct advantage as far as adaptation is concerned, especially in comparison with other species such as red beets and spinach. Soil temperature conditions required for germination are similar to those for red beets as are the mean temperature requirements for growth and quality. Moisture requirements for germination range from intermediate amounts to field capacity.

Production Regions

Most commercially produced chard is grown on a small scale and usually directed to nearby markets. California, New Jersey, Texas, and Florida are the main production regions. With the increased availability of these products, cooked-vegetables like chard and table beet tops are becoming popular.

Soil and Fertilizer Requirements

Chard has not yet come under the close scrutiny demanded of large scale commercial production aimed at maximum yield and quality. Most references to chard cite responses as good despite average soil type and fertility levels. A well-prepared, deep loamy soil high in organic matter will insure vigorous full-grown chard plants. The pH requirements are towards neutral-a range of 6.5-7.5 is highly satisfactory. The requirements for soil and fertilizers would approximate those given for table beets or spinach. Like table beets, chard is tolerant of Mg deficiency.

Cultural Practices

In the northern regions it may be necessary to start chard in a sheltered area, using cold frames, hot beds, or greenhouses and transplanting after the danger of frost is past, in order to gain an early start on the season. For the most part chard is direct seeded. The further south it is grown, the greater the possibility for sowing chard in the off season. Unless the roots and crown are adequately vernalized, the typical biennial habit of developing seed stalks may not occur. On the other hand, the second year's growth can be forced to vegetate again by removing the seed stalks as they emerge. Exposure to air pollutants like S02 and Pan may blemish and detract from the healthy appearance of the leaves. Due to the large leaf surface, chard should be provided with supplementary water between periods of rainfall to prevent serious water stress.

Pest Control

Local and/or manufacturer's recommendations for dealing with pests should be followed, and it should be determined which chemicals are registered for use on this crop.

Weeds: The recommendations for table beets should be followed as much as possible.

Insects and Diseases: The recommendations for table beet insect and disease control apply to chard.

Harvesting and Storage

Chard is hand harvested usually by removing only the outer, larger petioles approximately fifty to sixty days after planting. Care should be taken not to damage the growing tip. Stalks that are bright and blemish free are usually bundled together in bunches and quickly cooled to remove field heat and reduce the rate of respiration. Chard should be kept at a high relative humidity of 90 to 95 percent and held at O°C. Under these conditions chard can be stored up to ten days. Constant top icing is necessary to maintain shelf life.

CHAPTER 24

Tannia (*Xanthosoma sagittifolium* (L.) Schott)

Family: Araceae

Chromosome Number: 2n= 26

Origin: Two Araceae are attaining world importance as energy foods: the cocoyam, taro or dasheen (*Colocasia esculenta*) originating from Oceania and Southeast Asia, and the tannia, yautia or new cocoyam (*Xanthosoma sagittifolium*) from the American tropics.

Chemical Composition

Crude protein (4.33-5.92%), crude fiber (1.04-1.94%), crude fat (1.16-3.22%), ash (2.17-2.93%), with ede uhie exhibiting higher values than ede ocha. However, the carbohydrate and energy values were higher in ede ocha. For the mineral contents, the range values were: Ca (10.23- 41.17mg/100g), Na (15.22-17.66mg/100g), Mg (5.82-8.38mg/100g), P (12.31-16.19mg/100g), K (2.78-3.62mg/100g), with ede uhie showing higher values. On pH, TTA, starch and total sugar, it was observed that fermented ede ocha had higher pH value (6.82), TTA was higher in fermented ede ocha (0.92%), starch (24.66%) and total sugar (3.29%) values were higher in raw ede ocha than others. Anti-nutrient factors were generously reduced by processing methods especially boiling by between 50-100%, and fermentation by between 5-77% in most of the parameters analyzed. Functional properties ranged from BD (0.60-0.81g/mL), WAC (2.19- 2.71g/mL), OAC (1.45- 2.92g/mL), FC (16.38-19.52g/mL), FS (1.96-2.48 min/sec) and GT (60.85-81.05 0 C). While BD, WAC, OAC, FC were higher in the ede ocha flour, FS and GT were higher in ede uhie flour (Ukom and Okerue, 2005).

Botanical description

A herbaceous perennial, Xanthosoma sagittifolium has a corm or main underground stem in the form of a rhizome from which swollen secondary shoots, or cormels, sprout. Several large leaves also sprout from the main stem, which are sagittate and erect with long, ribbed petioles; inflorescences sprout between the leaves in a spadix, with a white 12 to 15 cm spathe which closes at its base in the form of a spherical chamber and opens at the top into a concave lamina; the spadix is cylindrical, slightly longer than the spathe, with female flowers on the lower portion, male flowers on the upper portion and sterile flowers in the middle portion. The spadices are rarely fertile and produce few viable seeds. The growth cycle lasts from nine to 11 months: during the first six months the corms and leaves develop; in the last four months, the foliage remains stable and, when it begins to dry, the plants are ready for the cormels to be harvested.

Climate and Soil Requirements

Tannia prefers warm humid climate prevalent in the tropical regions of the world. For sustained growth and production it requires an evenly distributed rainfall of about 2000 mm. The crop performs well in well-drained loamy soil with slightly acidic reaction (pH of 5.5 to 6.5). Tannia grows best as an up-land rain-fed crop on well drained soils where rainfall is well distributed throughout most of the year

Varieties and Hybrids

There are no specific released varieties of tannia in India. Only local selections with low acridity, good texture and suitability of leaves for making deep fried rolls are available. Where corms and carmel yields are being preferred, local collections with high yield and less acridity are desired. For making rolls, cultivars with high leaf yield and less acridity are preferred.

Corms and cormels are the usual planting material. Healthy cormels of bigger size and 20-25 cm long are commonly used. The setts from the top portion of the main corm with a thickness of 5-10 cm containing the apical bud are also used for propagation. In home gardens, it is usually grown as a ratoon crop. The cormels are separated annually from the main corm, which is covered with soil after removing the cormels. This practice can be repeated up to 5-6 years. In some parts of Maharashtra and Gujarat where only leaves are harvested, the main corms and cormels are left in the field. They continue to act as new planting material.

Field preparation could be done by ploughing in the summer before the onset of monsoon. This is followed by formation of ridges and furrows 90 cm apart. Planting of setts or cormels is done on the ridges at a spacing of 60 cm to 1m.

Mode of Pollination

Tannia seldom flowers, and some cultivars do not at all. When flowering does occur, the inflorescence arises in leaf axils, and depending on species, one or several species may form flowers, depending on the species are unisexual or bisexual. All are sessile and form on a spike (spadix) enclosed in large bract (spathe). In the unisex inflorescence, pistillate flowers are at the base of the spadix, staminate flowers at the top, and in between are a grouping of abortive sterile flowers. Insects pollinate the flowers, but fruit and seed set are rare. Fruits are berries, each containing 2-5 small (1 mm), oval, hard and difficult to germinate seeds.

Management of water, nutrients and weeds

Irrigation: Tannia prefers upland rainfed conditions more than low land water logged conditions. However, sufficient water needs to be provided for maximum vegetative growth and leaf production during the starch storage period. Under rainfed situation, a short-term drought often may accelerate crop maturation.

Nutrients: Application of manures and fertilizers at the rate of 12.5 t ha^{-1} of FYM, and 80:60:80 kg ha^{-1} of N, P_2O_5 and K_2O is recommended. Traditionally full dose of phosphorous and half dose of nitrogen and potash are applied as basal a week after sprouting and the remaining half dose, a month after the first.

Weeding: Two weedings and earthing up operations have been suggested at 30th and 60th days after planting. Application of mulch soon after planting is beneficial to the crop as it conserves soil moisture, reduces temperature around the sett and also suppresses the weeds.

Harvesting and Yield

Tannia is usually hand-harvested by digging out mature corms. The crop is ready for harvest within 9 -10 months after planting with expected yields between 25 and 30 t ha^{-1}. If the crop is grown for leaf purpose, then harvesting of the leaves should be done immediately after they attain maturity. A leaf size of 20-30cm x 20-30cm is ideal for making roll. Thus 40-50 leaves can be harvested from 1 clump during its entire growth period. Leaves are harvested along with the petiole. Petioles are also used as a vegetable. If the crop is intended for taking corms, then it can be harvested when older leaves start turning yellow. Its corms do not deteriorate if left unharvested in the soil and in this way harvesting can be done from 6 months onward up to 9 months. If it is to be grown as a ratoon crop, the soil around the clump should be dug to remove the mature cormels. The main corm is retained along with roots and covered with soil. This can be practised for 4-5 years.

Post harvest Handling

The corms can be stored for a period of 4-5 months under well ventilated and dry conditions. Care should be taken to avoid injury to the corms while harvesting. After harvesting, curing of the corms and cormels is done in sunlight for 4-5 days. The corms and cormels can be stored embedded in dry soil or sand for a period of 4-5 months under ventilated, dry and semi-dark conditions.

Diseases and Pests

No major disease has been reported for this crop. The insect pests of this crop are similar to other aroids and yams as describe in yam.

CHAPTER 25

Yam (*Dioscorea spp.*)

Family: Dioscoreaceae

Dioscorea is the largest genus of the family, comprising approximately 600 species (Pedralli, 2002), of which 10 species (*D. cayenensis* Lam., *D. rotundata* Poir., *D. alata* L., *D. bulbifera* L., *D. esculenta* (Lour.) Burkill, *D. trifida* L., *D. opposita* Thunb., *D. nummularia* Lam., *D. transversa* R. Br. and *D. pentaphylla* L.) are edible and represent the basic food source for more than 100 million people worldwide, especially in the humid and sub-humid tropics (FAO, 2009; Lebot, 2009). The first three species mentioned above are of great importance in the world, especially in Africa, where, along with cassava (*Manihot esculenta* Crantz), they represent the primary source of carbohydrates among different communities and ethnic groups (Odebunmi et al., 2007).

Botanical Description:

Vigorously twining herbaceous vine, from massive underground tuber. Stems to 10 m (30 ft) or more in length, freely branching above; internodes square in cross section, with corners compressed into "wings," these often red-purple tinged. Aerial tubers (bulbils) formed in leaf axils (not as freely as in D. bulbifera), elongate, to 10 cm (4 in) x 3 cm (1.2 in), with rough, bumpy surfaces. Leaves long petioled, opposite (often with only 1 leaf persistent); blades to 20 cm (8 in) or more long, narrowly heart shaped, with basal lobes often angular. Flowers small, occasional, male and female arising from leaf axils on separate plants (i.e., a dioecious species), male flowers in panicles to 30 cm (1 ft) long, female flowers in smaller spikes. Fruit a 3-parted capsule; seeds winged.

Climate and Soil

Yams thrive well under warm sunny weather (temperature of 25 to 30 °C) with plenty of rains. Adequate moisture is required for the growth, development and yield of yams. Evenly distributed annual rainfall of 100 -150 cm spread over six to seven months period is ideal for *D. rotundata*. However, *D. esculenta* comes up well in places receiving a rainfall of 90 to 100 cm. .*D. cayenensis* requires fairly long rainy season of about 10 months each year. On the other hand, most *D. rotundata* cultivars can be grown with far less rainfall and seven months of wet weather is often sufficient. The most critical period of rainfall/moisture is the first five months after planting. Yam production is also significantly influenced by day length. It has been found that shorter days favour tuber formation and development while longer days (> 12 hours) apparently influences vine growth.

Deep and loose friable soil with adequate nutrient is well suited for the growth and development of yams. It is also essential to have proper drainage facility in soil, as stagnant water is likely to damage the tuber.

Cultivation Practices

Yams are commonly propagated vegetatively. Either tuber pieces or small whole tubers are used. Propagation by vine-cuttings is not advisable because of slow tuber production. A seed tuber weighing 125-150 g is optimum size for planting in *D.esculenta* (Chiness yam), while in *D. alata* (Water yam) and *D. rotundata* (White yam), 200-250 g tuber pieces or whole tubers are used. Usually larger whole tuber is cut into pieces (setts) consisting of top, middle and bottom. Seed tubers and top portions are ideal for planting. Before planting tuber pieces are dipped in cow dung slurry and dried in shade to protect from damage. Drying of cut pieces gives healing effect, encouraging callus formation. Yam tubers have a long dormancy period (2-3 months depending on species). The degree of dormancy decreases during storage resulting in sprouting and weight loss, a major problem in yam cultivation. To maintain dormancy and controlling storage sprouting, soaking freshly harvested tubers of lesser yam and greater yam in 0.1 % solution of maleic hydrazide for 10 hr is effective. Dormancy can be broken by dipping of tubers quickly in 4-8% solution of ethylene chlorohydrin followed by dry storage. Adequate precaution needs to be taken during this treatment.

In India, traditionally yams are grown in homesteads. The package of practices consists of ploughing or digging of the land up to a depth of 15 -20 cm followed by opening of pits of the size 45 x 45 x 45 cm. About 3/4 th size of these pits are to be filled with dry farm yard manure and top soil. Seed yams or yam setts are then planted in it and mulched with green or dry leaves and soil heaped over them.

Maximum tuber size is obtained when plants are 2 meters apart. Closer spacing results in decreased tuber size but produces more yield per unit area. It is suggested that wider spacing should be used in heavy soils and if staking is not intended. Under ridge and furrow system, 70 cm spacing between ridges and 45 to 60 cm spacing between plants so as to accomodate 25,000 plants ha-1 is generally adopted. A spacing of 1 x 1 m with a plant population of 10,000 plants ha-1 for *D.alata* and *D.rotundata* has been recommended. In general for dwarf bushy plants, a close spacing of 60 x 60 cm is optimum and economically viable.

Mode of Pollination

Yams are vegetatively propagated dioecious species. In *D. alata,* clonal selection had been the only mode of improvement as flowering is erratic and many of the clones are sterile. In the available germplasm flowering of clones in any year is not assured. Moreover, synchronous flowering of male and female clones necessary for sexual reproduction is very rare. It was also found that a large number of female clones are sterile.

In *D. esculenta,* flowering has been found to be very rare. Only few accessions in the germplasm had been found to flower. The flowering types are sterile males and females are not observed among cultivated types. Hence, with the available stocks, genetic improvement of the species does not seem possible in the immediate future. As in the case of *D. alata* in which sexual reproduction was found possible only recently, it is hoped that in *D. esculenta* also fertile types could be located sooner or later.

Varieties and Hybrids

A number of varieties have been released in yams. The improved varieties show a yield potential of 18-25 tonnes/ha in greater yam and 18-30 tonnes/ha in lesser yam, while in white yam it is 33-40 tonnes/ha. Important varieties are Sree Keerthi Sree Roopa for Greater yam Sree Latha, Sree Kala, Konkan Kanchan for Lesser yam and Sree and Priya Sree Subhra for White yam *D. rotundata* is found to flower freely and profusely under Kerala condition. Unlike the other edible yams, the clones are highly fertile and seed setting under open pollinated condition is profuse. The crossability of this fertile species as a result of synchronous and profuse flowering of male and female clones is providing the tool for further improvement of this exotic species.

Irrigation:

Yams are basically tropical crop and hence prefers adequate moisture and plenty of sunshine. Irrigation is seldom given to yam crop in

India, especially under the Kerala climate as the state receives plenty of rains. In most yam growing countries of the world, it is grown as a rainfed crop and therefore, irrigation is not a well defined cultural practice in yams.

Nutrient:

Yams are reported to be highly efficient in utilization of native and applied nutrients in soils. For each metric ton of dry matter produced, yam tubers removed almost the same quantity of nutrient from the soil as that of potato crop, but about four times as much N and twice as much P and K compared to the cassava crop.

Yams are highly responsive to manures and fertilizers. Soils containing less than 0.1 per cent N, 10 ppm P and 0.15 me/100 g exchangeable K evoked high yield response in yams. Manuring at the time of ridging or mounding has been found to be beneficial. Apart from farmyard manure (FYM), compost, green manure, etc., are commonly used as organic manure. In the early stages of growth, the newly developing shoot is sustained by the food reserves of the tuber and hence no additional nutrients are required during the first few weeks of growth. Later as the plants divert from foliage production to tuber production, potassium becomes the limiting element. Yams are highly efficient in extracting phosphorous from the soil and seldom need additional requirements partly due to the association of mycorrhizal fungi, which help yam roots to absorb P effectively from the soil. Fertilizer application to yam crop in India is still not widely practised because the crop is usually raised in the homesteads where plenty of organic wastes along with cattle manure and ash are available. In general yams respond well to nirogen and potassium. The general recommendation however, is 80:60:80 kg ha^{-1} of N,P_2O_5 and K_2O in two split doses in addition to basal application of FYM @ 10 -15 t ha^{-1}. Optimum time of application of the first dose of fertilizers is one month after planting when the plant changes from dependence on the sett to true autotrophy. This helps in development of large leaf area, thus providing high photosynthetic ability for tuber development. Second split application of fertilizers should essentially coincide with the tuber bulking stage, when the demand for plant nutrients would be at its peak, which has been reported as between 3 and 5 months after planting. Application of half dose of N and K and full dose of P at the first application and the remaining half of N and K at the second application is generally advocated. Amending acid soils with lime to a minimum of pH 5.5 produces higher tuber yield.

Weeds:

The weed control measures should begin at an early phase of the plant growth as the plant is particularly sensitive to competition of weeds during this stage. The critical stages for weed interference in yam synchronises with leaf development and tuber bulking. Traditionally yam crop is weeded manually by using hoe, as it is the most effective and cheaper method as compared to mechanical and chemical devices. Application of a pre-emergence herbicide like Diuron @ 3 -3.5 kg ha^{-1} has been found to be quite effective Another crucial intercultural operation in yam cultivation is earthing up which usually accompanies weeding and fertilizer application. Earthing up is an essential feature because it makes up for the loss of soil from the mounds due to run off. As result of earthing up, exposure of the bulking tuber is prevented.

Harvesting and Yield

Yam crop reaches senescence between 7 and 9 months after planting. Time of harvest is not a critical criterion in yams and hence it could be done a couple of weeks before senescence sets in or to the extent of one or two months after senescence. In single harvest, 7-8 months of crop growth after planting are needed for *D.esculenta* while 8 -9 months are required for D.alata and *D. rotundata*. Yams are harvested 7-9 months after planting. The leaves turn yellow and vines start drying up at maturity. Delay in harvesting up to 2 months does not affect yield. Lesser yam yields 25-30 tonnes/ha, while Greater yam yields 33-40 tonnes/ha, whereas white yam yields 18-25 tonnes/ha.

In India, yams are harvested during December-January. Yams need proper storage for 2-3 months. The main problems in storage are rotting and sprouting of tubers resulting in storage loss. The common practice for controlling rotting and sprouting are to remove the rotten tubers and breaking off the sprouts periodically. Sprouting can also be controlled by soaking tubers in 0.1 % maleic hydrazide solution for 10 hr. As the tubers are stored in open barn, the weight loss and rotting in tubers are more. But covered barn structure is more effective in reducing weight loss. Weight loss also can be minimized by heaping tubers inside sand and dry soil treatment.

Post harvest Handling

Yam tubers in storage are subject to considerable loss of weight due to respiration, dehydration and sprouting. There are several methods of storing fresh yams: by tying them on vertical stakes in a shaded or un-shaded area, by heaping them pyramid-wise in houses or under shade or by placing them in pits. A 66 yam barn is a shed with woven stick-walls and a thatched roof or open shaded area with proper fencing. In the barn, the individual tubers are tied to walls. Yams could be safely stored up to four months by spreading them on racks under a thatched room with ample facilities for free flow of air and diffused light.

Dormancy period in yams lasts for 2 -3 months after harvest, but some yam species like *D.rotundata* have longer period of tuber dormancy as compared to *D. cayenensis*. A low level of glutathione is probably associated with prolonged tuber dormancy. Dipping of yam tubers in 2 -chloroethanol of 2 -8 per cent solution has been reported to be effective in breaking dormancy and enhancing sprouting. Soaking the freshly harvested tubers of *D.esculenta* and *D. rotundata* in 1000 ppm solution of maleic hydrazide for 10 hours was very effective in controlling storage sprouting.

Diseases and pests

Leafblight/ anthracnose (*Colletotrichum gloeosporiodes, Glomerella cingulata*)

Brown spots encircled with yellow halo develop at the apex and edge of leaf. Light and dark brown concentric circles are formed. Leaves are damaged, reducing yield potential. Fungicides like spray of Mancozeb (0.2%) is recommended for its control.

Leaf spots (*Cercospora carbonaceae C. dioscoreae*) :

Small brown to black spots, enlarge and coalesce together. In severe cases entire field is seen as burnt and blighted. Leaves are damaged. Spray of Mancozeb or Captan (50wp) (0.25%) as per need is advocated for its control.

Mosaic (Yam mosaic virus) :

Leaf mottling, chlorosis, cupping, puckering, veinbanding and necrosis are symptoms. Leaves are affected, reducing yield potential. Use of disease-free planting material is recommended for managing or reducing the disease incidence.

CHAPTER 26

Yam Bean (*Pachyrrhizus erosus* (L) Urban)

Family: Leguminaceae

Chromosome Number: 2n= 22

Origin: Africa

Botanical Description

The morphological characteristics delimitating the genus are the presence of a 'falsebeard' of short hairs along the dorsal (adaxial) side of the ovary, continuing almostto the base of the stigma along the incurved side of the style. The stigma itself issubglobular and is placed almost terminally. Also, all species have tuberized roots. A hairy twining herbaceous plant, woody at the base, trailing or climbing to about 6 m. The leaves are alternate, trifoliate, with petioles 3-18 cm long and ovate or rhomboidal leaflets which are toothed or lobed, about as broad as long, usually large, in the range 4-20 cm. The flowers are in long axillary racemes, 1-5 borne in each of several clusters along the peduncle; the petals are violet or white, 1.5-2 cm long and broad. The pods are 7.5-15 cm long and about 1.5 cm broad, flattened, almost smooth at maturity, containing 4-12 seeds which are yellow, brown or red, almost square and flattened, 5-10 mm in diameter. Tuberous roots, frequently turnip-shaped, are borne at the base of the stem, and may be solitary or several, simple or compound; normally they are about 10-15 cm in diameter. This species shows considerable genetic variability, and phenotypic studies have shown significant negative correlation in some genotypes between root yield and days to flowering, days to pod maturity and length of main stem.

Importance

Yam bean (*Pachyrrhizus erosus* (L) Urban) belongs to the family Leguminaceae and sub family Fabaceae (Papilionaceae). It is also called 'potato bean' in English and *Mishrikand* in Hindi. It is popular by known as 'Shankh alu' or 'Sankesh alu' in West Bengal, Assam and Orissa. Yam bean is a starchy root crop with comparatively high sugar content and a moderately good source of ascorbic acid. Tubers contain more than 82% water, 1.5% protein, 10% starch and

5-6% sugar. It is good for salad preparation and in many countries, the seeds are however poisonous. Since Yam bean is a leguminous tuber crop its cultivation, like other legumes, helps improve the soil fertility.

The crop is supposed to have originated in hot moist basin of the river Amazon and has been under cultivation in Mexico and South America from pre-Colombian period. The crop is now cultivated in the Philippines, China, Indonesia, Nepal, Bhutan, Burma and India. In India, it is grown in parts of West Bengal, Bihar, Orissa and Assam.

Climate and Soil Requirements

Yam bean requires a hot humid climate and adapts well in sub-tropical and hot temperate zones. The basic requirement is frost-free condition during the growth period. It is grown up to an altitude of 1000 m. It has been observed that thermo-periodism has got a definite effect on tuberization. Though yam bean requires 14-15 hours of photoperiod for good vegetative growth, short days with cool nights are effective for good tuberization. A well- Excessive rain causing water logging is deleterious to the crop. Cool climate during early growth period adversely affects the initiation of tubers and also results in a prolonged vegetative phase.

Fertile, well-drained, sandy loam soil is best suited for cultivation of yam bean. This crop adapts well to loamy and clay loam soil. Water logging adversely affects yam bean cultivation. Optimum soil pH requirement is 6.0 -7.0.

Varieties and Hybrids

There are 2 types of cultivars available in market. Mexican types have larger tubers (500- 700g), whereas local types have smaller tubers (200-300g). The Mexican types are less sweet compared to the local ones. The Mexican types have a tendency to develop cracks on their tubers. They are less preferred in the Indian markets. The flesh is white with less fibre. There is no cracking. Rajendra Mishrikand 1, an improved variety, is very popular in North Bihar. It is recommended for cultivation in Bihar and West Bengal. Its average yield is 40-55 tonnes/ha in 110-140 days. Other promising line, L 19, gives better yield in Bihar, West Bengal and Orissa.

Cultivation Practices

Yam bean is propagated usually through seeds. The seedpods are generally 7 –15 cm long with 8-10 brownish-yellow to red seeds.

Seeds are dorsi ventrally flat. Plants can also be raised from sprouted roots of previous crop but this is not normally practised. The seed rate varies according to spacing of the crop. A seed rate of 20-60 kg/ha is generally adopted by the farmers. The mature pods containing seeds are source of a toxic substance 'rotenone' and are sometimes harmful for cattle grazing in the field.

Deep ploughing of land twice using a mouldboard plough is essential. Plank the soil after each ploughing to have a well pulverised soil as well as to conserve moisture. A good tilth is required for yam bean cultivation. Traditionally Yam bean is sown during June-July with the onset of rain in Northeastern India. Yam bean seeds can be sown on hills at the rate of 3-5 seeds per hill. Hills are prepared at a spacing of 0.75 -1.00 m with 15 cm height. It has been observed that planting the seeds on ridges results in better yield.

Normally yam bean starts flowering 75 days after sowing. It is desirable to remove the flowers and not allow the plant to bear pods for getting better tuber yield. It has been observed that spraying 2,4-D (50 ppm) at flower initiation stage causes abscission of flowers and results in better yield of tubers.

Mode of Pollination

Yam bean is a self pollinated crop. The flowers are violet or white and borne on erect pedicels in racemes producing 7-14cm long and 1-2 cm wide pods. Flowering commences in the shorts days when new vegetative growth diminishes and storage root enlargements accelerates. Normally 10 months are required to produce mature seeds. Cultivars with greenish-brown colored seeds are preferred because they are more productive than those with either green or brown seed.

Irrigation: Normally there is no need to irrigate the *kharif* crop. In case there is scarcity of rains, irrigation is required. For September sown crop, it is advisable to give supplementary irrigation so that the crop will not face moisture stress during tuberisation.

Nutrients: Apply well rotten compost or farmyard manure (FYM) at the time of land preparation. Application of 15-20 tonnes of FYM or compost along with about 80:40:80 kg N, P_2O_5 and K_2O ha^{-1} is recommended. Entire dose of P and K may be applied as basal dose at the time of planting along with half dose of nitrogen. Remaining half dose of N is top-dressed at 40-50 days after sowing along with interculturing and earthing up.

Weeds: Weed infestation is more in *kharif* crop compared to the late (September-sown) crop. It is advisable to do the first interculturing 40 days after sowing and add the remaining half dose of nitrogen and do earthing up. Second weeding is to be done 30 days after the first weeding. The field has to be kept free from weeds by manual weeding or using mechanical implements.

Harvesting and Yield

Monitor the development of the tubers. They are suitable for harvest when they reach 10 cm across but can be left till double that size in most varieties. If left too long or if the vines flower heavily the tuber quality will reduce. Yam bean normally takes 150 days after sowing. Usually it is harvested on the occasion of 'Saraswathi Pooja'. If harvesting is delayed, chances of cracking of tubers are more. Shallow irrigation may be given just before digging the tubers manually. The above ground portions are trimmed before digging out the tubers. Harvested tubers can be stored for 2-3 days without any deterioration. If the tubers are stored for a longer period, the creamy colour of the skin changes to purplish brown and loses water resulting in reduction in weight. The harvest can be adjusted to the demand by leaving the crop in the soil without removing top portion. The average yield of local cultivars is 18-20 t ha^{-1} while that of improved varieties like Rajendra Mishrikand is 36-40 t ha^{-1}.

Post harvest Handling

It can be harvested early or late according to market demand. It is possible to harvest the crop with smaller-sized tubers after 100 days. Otherwise it can be left in the field up to 140 days for better size. Traditionally, the trend is to harvest the crop on the occasion of' Saraswati Puja' (January) with the start of spring season because of market demand. Delayed harvest leads to fibrous flesh along with cracks in tubers leading to deterioration in tuber quality. Care should be taken to avoid cuts and bruises on tubers.

Tubers are graded according to their size. They can be stored for 3-5 days without any deterioration. When the tubers lose their weight, they become sweeter. Since tubers are firm, no specialized packing is required. They are packed in jute bags and brought to market.

Diseases and Pests

No major disease has been reported for this crop. The pests of this crop are similar to that of aroids and yams as described under yam.

CHAPTER 27

Winged bean (*Psophocarpus tetragonolobus* L)

Family: Leguminoseae

Chromosome number: 2n= 18

The plant is twining, glabrous herb, but usually grown as annual. The green pods, immature seeds, young leaves, flower and tubers are eaten as vegetable. In India, it is commonly known as Goa bean. The winged bean is exceptional among all food plants because practically all parts are edible and is produced at every stage of the life cycle.

Botany

The winged bean plant grows as a vine with climbing stems and leaves, 3–4 m in height. It is an herbaceous perennial, but can be grown as an annual. It is generally taller and notably larger than the common bean. The bean pod is typically 15–22 cm (6–8.5 in) long and has four wings with frilly edges running lengthwise. The skin is waxy and the flesh partially translucent in the young pods. When the pod is fully ripe, it turns an ash-brown color and splits open to release the seeds. The large flower is a pale blue. The beans themselves are similar to soybeans in both use and nutritional content (being 29.8% to 39% protein). The appearance of the winged bean varies abundantly. The shape of its leaves ranges from ovate to deltoid, ovate-lanceolate, lanceolate, and long lanceolate. The green tone of the leaves also varies. The stem, most commonly, is green, but sometimes boasts purple.

The pods are usually rectangular, but sometimes appear flat. Pods may be coloured cream, green, pink, or purple. Pods may be smooth are appear smooth or rough, depending on the genotype. Seed shape is often round; oval and rectangular seeds also occur. Seeds may appear white, cream, dark tan, or brown, depending on growing and storage conditions.

Use:

The young pods are succulent and eaten as cooked vegetable. They contain appreciable amount of calcium, iron, thiamine and ascorbic acid. Mature pods become tough and fibrous. The tuberous roots are eaten as a delicacy. Dried roots contain moisture 9.9%, protein 24.6%, carbohydrates 56.15, fat 1%, and ash 3.9%. The starch grains of tubers vary in size and shapes. The young leaves are also eaten as a vegetable.

Origin and distribution:

It is believed to be native of either India or Mauritius or originated in the African region where all the 4- 5 species of this genus are found wild. It is a perennial climber with tuberous roots cultivated in India, Burma, tropical Africa, Asia and Madagascar for it's edible pod and tuberous roots. In India, it is cultivated in Maharashtra, Bihar, Orissa, West Bengal, Assam, Tamil Nadu and Kerala.

Climate and soil;

Since most varieties of winged beans need short days to initiate flowering, yet are frost sensitive, they must be grown during the winter in South Florida while the days are short. Farther north, the shorter days of late fall give best results. Seeds for planting are not easy to find, although some seed catalogs advertise them. Plants grow best under hot, wet conditions and grow best with 60 to 100 inches of rainfall or irrigation per year. It prefers comparatively warm and humid regions. The optimum temperature for tuberization is 16- 25 °C. The high temperature promotes vegetative growth, but inhibits tuberization.

It thrives best in light, well drained, loamy soil and warm humid region with partial shade. It can be grown as a green manure cover crop, fodder crop and restorative fallow crop because of its nodulation.

Seed rate:

The seeds are sown in early rainy season. Generally 50- 60 kg of seeds are sufficient for planting one hectare land.

Sowing time:

Best results under north Indian conditions can be obtained in terms of grain yield (15q/ha) when crop is planted during May-June. Genotypic differences due to photoperiod/temperature response exist. Long days can inhibit tuber initiation but may improve nodulation. In India, and in many other tropical countries, the onset of rainy season is preferred for raising this crop. However, in the highlands of Papua New Guinea, it is grown (in drier season because fewer pests and diseases occur during this period.

Spacing: In India, experiments on spacing (Mishra and Mishra, 1980) showed that 90 cm spacing bothways i. e. inter and intra row distance is quite suitable.

Cultural practices:

Earthing up around the plant after 3 months of planting will induce tuber formation.

Harvesting:

Winged bean flowers during periods with short day lengths. Pods are harvested when still tender, about 2 weeks after pollination. The pods are harvested before they mature. Each vine yield about 25 pods within every 5- 6 days and will continue if the vines are manure every 2 or 3 weeks. The yield of tubers are very from 2.5- 6 tones/ ha.

CHAPTER 28

EXPLORATION AND UTILIZATION OF INDIGENOUS VEGETABLE CROPS

Indian gene centre has rich biodiversity in vegetable crops. Vegetables are grown from dry temperate to humid tropic between the altitudes from sea level to snow line. Nowadays, vegetables are more popular considering their nutritional and pharmaceutical importance. Presently more attention is being focused on the introduction and production of indigenous vegetables. Yet the region is rich resources and underutilized plant species with potential food value, about which there is little knowledge. The underutilized plant species of economic importance are the key for sustainable agriculture in most of the developing countries facing resource constraints as well as rapid depletion of natural resources due to increasing population pressure. Poverty at both rural as well as urban areas leads to various health and nutritional problems, respectively.

These crops could also help in poverty alleviation by providing income generating opportunities to the farmers. But many of these crops are in danger of loss through competition resulting from the introduction of higher value commodities into the farming system. Underutilized crop species have a potential that needs to be exploited to advantage. Most of these crop species do not require high inputs, can be grown in marginal and degraded lands, and at same time contribute to increased agricultural production, crop diversification and a better environment. In addition, such efforts will be help in conserving and using the genetic resources of underutilized vegetable species. Future strategies of the underutilized vegetable crops and their importance among the growers and consumers developing improved varieties, having wider adaptability and developing suitable cultural practices for different crops ensuring timely input supply, creating indigenous market.

Exploration:-

Exploration is trips for collection of various forms of crop plants and their related species. it is the primary source of all germplasm present in various germplasm collection.

Objectives:-

It is the collection of germplasm needed by the breeders involves the collection of variability present in crop plants and their wild relatives. It provides access to the materials of special interest. e.g. new genes, new species etc.

Indigenous Vegetable:-

Indigenous vegetables are domesticated or semi -wild vegetable crops grown in particular regions as an integral part of a local food system. Or Indigenous vegetables are those crops which are grown primarily in their centre of origin or centre of diversity by traditional farmers, where they are still important for the subsistence of local communities. These plants are evolving naturally, and then adapt over time to their new homes.

Why they are underutilized?

These species as underutilized since past they were grown more widely or intensively but are now falling into disuse for a variety of agronomic, genetic, economic and cultural reasons. Some species may be widely distributed around the world but tend to occupy special niches in the local ecology and in local production and consumption systems. Farmers and consumers are using these species less because they are not competitive with other species in the same agricultural environment.

Characterization of Indigenous Vegetable Crops

1. Require only limited external inputs for production
2. Suitable for organic production
3. Suitable for cultivation on marginal land (poor soil fertility)
4. Suitable for stabilization of fragile ecosystems
5. Fit into small-scale farming systems
6. Possess traditional, local and regional importance
7. Easy to store and process by resource-poor communities
8. Market opportunities available
9. Possess high nutritional and/or medicinal value

Minor vegetables:-

Crop	Botanical name	Edible part
Agathi	*Sesbania grandiflora*	leaf
Amaranthus	*Amaranthus spp*	Leaf
Basella	*Basella spp.*	Tender leaves , stem
Cassia	*Cassia alata*	Tuber, leaves
Celeriac	*Apium graveolens*	Swollen stem base
Chekurmanis	*Sauropus androgynus*	Stocky head of leaves
Chive	*Allium schoenoprasum*	Bulbs
Chive	*Allium schoenoprasum*	Bulbs
Chow-chow	*Sechium edule*	Fruit. stem, tender leaves
Coleus	*Selenostemon rotundifolius*	Corn
Curry leaf	*Murraya koenigii*	leaves
Endive	*Chichorium endivia*	Stocky head of leaves
Globe artichoke	*Cynara scolymus*	Bud scales, flower head
Ivy gourd	*Coccinia grandis*	Fruit
Leek	*Allium ampeloprasum*	Leaves, blanched stem
Lettuce	*Lactuca sativa*	Leaves
Moringa	*Moringa oleifera*	Pod
Parsley	*Petroselinum crispum*	Leaves ,root
Spine gourd	*Momordica dioica*	Tender green fruit
Taro	*Colocasia esculanta*	Corms
Water chestnut	*Eleocharis dulcis*	Tubers
Winged bean	*Psopbocarpus tetragonololobus*	Pod, seed
Yam	*Dioscorea spp.*	Tuber

PASSPORT DATA

Passport data are information about a germplasm sample and the collecting site recorded at the time of collecting or plant germplasm exploration.

1. Date….,
2. Accession No….. ,
3. Common Name….,
4. Name of farmer..,
5. District ….,
6. Latitude ….,

7. Altitude.... ,
8. Rainfall ....,
9. Temperature...
10. Longitude....,
11. State....,
12. Village/Block....,
13. Cultivar Name....,
14. Species Name....,
15. Collector No. ....,

Germplasm Utilization:-

Germplasm can be used in breeding program in the following 3 ways:

1. It may be direct release of variety.
2. It may be subjected to selection for developing a variety
3. It may be used as parent in hybridization programme

It is two type;

a. **Intra-specific:-** Intra-specific hybridization refers to crossing of two different species of same genus.

b. **Inter-generic:-** inter-generic hybridization refers to crossing of two different genera of same family.

Evaluation: - The newly selected lines/strains/populations are tested for yield and other trait and their performance is compared with the existing best varieties. It is two types abiotic and biotic

Abiotic:-Drought, heat tolerance, Salinity.

Nutrition Evaluation:- Protein, TSS, Fiber, Lycopene, Carotenoids, Capsaicin, Vitamin C, Sugar, Dry matter, Anti-oxidant, Acidity, Colour value, Vitamin-A , pH.

Biotic:- Virus resistance, Diseases resistance, Insect resistance.

GERMPLASM CONSERVATION

Germplasm has to be maintained in such a state that there is minimum risk for its loss and that either it can be planted directly in the field or it can be conserved for planting with relative ease.

IN-SITU CONSERVATION: This technique involves the maintenance of genetic variation at natural habitat or in the area where it grows natural. This is achieved by protecting this area from human interference. In situ conservation can be done in: National park Biosphere reserves on farm community efforts.

EX-SITU CONSERVATION: Conservation of germplasm away from its natural habitats is called as ex- situ conservation. It can be divided in 4 ways:

1) Seed gene bank.
2) Plant / field gene bank.
3) Chryo gene bank.
4) DNA bank.

An approach to assess the yield potential and ecological requirement of underutilized crop:- Field

1) Germplasm selection
2) Prcliminary cvaluation
3) Characterisation of selected germplasm
4) Agronomic studies On-station field experiments
5) Germplasm collection Agro-ecological zoning
6) Recommended agronomic practices

Organizations -Underutilized vegetables crops:-

1) International Plant Genetic Resources Institute (IPGRI), Rome, Italy.
2) Conservation and use of PGR of Neglected and Underutilized crops, Africa
3) Indigenous knowledge, gender and nutrition of leafy green vegetables, Africa
4) Home garden conservation of underutilized and neglected crops, America
5) Food and Agriculture Organization (FAO), Rome, Italy
6) AVRDC- Tapei, Taiwan

INDIGENOUS VEGETABLES

IVY GOURD

Coccinia indica L.

Common Name: Little gourd or kundru

Family: Cucurbitaceae

Distributed China, Tropical Africa, Malaysia India: Chattisgarh, West Bengal, Bihar, Orissa Karnataka

Cultivar: Twenty accession collected from Orissa and isolated an elite genotype namely CHGI 15 and CHGI 13.

Uses: Dried leaves are a useful treatment for diabetes.

Propagation- By Vine/stem cutting.

Harvesting: 50-60 days after planting

Yield: 15-20 kg/ vine/ year

SPINE GOURD

Momordica dioica.

Common name: Kakrol

Family: Cucurbitaceae

Distribution: India, Pakistan, Bangladesh, Sri lanka

In India; Punjab ,Uttar Pradesh, Rajasthan, Madhya Pradesh, Kerala and Maharashtra

Cultivar: IIVR collected genotype from Mirzapur (U.P) -MS-15, MS-18 -Indra kankad- released by Indira Gandhi Agri. University, Ambica (Chattisgarh)

Uses: Immature and lender fruits are cooked as vegetable. Fruits are also used for pickle preparation. It have several medicinal and curative properties.

Propagation- stems cutting, seed.

Harvesting: 70-80 days after planting. Yield: 10-12 q/ ha

COLEUS

Coleus rotundifolius Benth.

Common name: koorka/ Ratala

Family: Labiatae

Distribution: Indonasia, Sri lanka, South Africa. In India; Kerala, Tamil Nadu, Andhra Pradesh

Varieties: 1.Shredhara released by CTCRI 2.Nidhi released by Kerala Agriculture University.

Uses: Usually cooked as a curry along with various spices and eaten with rice. They can also boiled baked or fried similar to potato.

Propagation- vine cutting and

Seed harvesting: five months after planting , Yield:10-12 t/ha

CURRY LEAVES

Murraya koenigii L.

Common name: Kari patta

Family: Rutaceae

Distribution: Burma , Srilanka, Pakistan, Bangaladesh. In India: North India, Assam, Kerala, Sikkim etc.

Varieties: DWD-1 (Suwasini) DW-2 (Dharwal dwarf -2)

Nutritional value: Nutrient composition (g or mg/ 1oog), Carbohydrates (g) 11.6, Protein (g) 6.8 Calcium (mg) 570.0 Phosphorus (mg) 200.0 Carotene (µg) 5706.0

Uses: The leaves use for Ayurvedic medicine, an anti-diabetic, anti-microbial, they also contain iron and most commonly used in curries,

Propogation: by seeds harvesting: the leaves should be plucked 30-35 days after sowing.

CHEKKURMANIS

Sauropus androgynus Merr.

Family: Euphorbiaceae

Distribution: Indonesia, Malaysia, Singapore. In India: North-East Himalayas, Tamilnadu

Varieties: no distinct type's varieties are available.

Uses: Composition (g or mg (µg) / 100g nutrient, Carbohydrates (g) 11.6, Protein (g) 6.8, Calcium (mg) 570.0, Phosphorus (mg) 200.0, Carotene (µg) 5706.0

Uses: Leaves are highly nutritious are called multivitamin which contain good amount of Vit. A, B, C and K .

Propagation: by seed, stem cuttings

Harvesting: 3-4 months after planting

Yield: average annual yield 30-35 t per ha.

SNAKE GOURD

Trichosanthes cumarina

Family: cucurbitaceae

Origin: India

Edible part: Fruits

Uses: as vegetable, salad and curry dishes etc

Propagation: by seeds

Harvesting: 3-4 months after sowing.

SWORD BEAN

Family: Fabaceae

Origin: Southern Asia or Africa.

Edible part young pods, stem tips & tender leaves.

Uses: Young pods, stem tips & tender leaves are used as cooked vegetable.

It is useful as a cover crop & also as green manure.

Propagation: by seeds

Harvesting: 3-4 months after sowing

WATER CONVOLVULUS

Ipomcea aquatica

(Syn. Ipomoea reptans)

Family: Convolvulaceae

Origin: Africa.

Uses: The tender shoot tips and leaves are used as green leafy vegetables. Foliage stems are excellent sources of vitamin A.

Propagation: Stem cutting or seeds.

Harvesting: Shoot tips are harvested 30-40 days after planting of cutting and about 60 days after seed sowing.

JACK BEAN

Canvalia ensiformis

Family: Fabaceae

Origin: Native of Central America & West Indies

Edible part: Young pods & immature seeds

Uses: It is a good source of protein Improves soil fertility climate Tropical and Sub-tropical.

Propagation: Seeds sowing time: June–July yield 3–4 t/ha.

ALOCASIA (GIANT TARO)

Alocasia macrorrhiza

Family: Araceae.

Origin: Sri Lanka distribution India, Malaysia, Indonesia edible **part:** corm

Uses: Corms are used as vegetable. It is good source of carbohydrate.

Propagation: By corm

Harvesting: Ready to harvest 12-18 months after planting.

Yield: Average yield is 30 t/ha.

WINGED BEAN

Psophocarpus tetragonolobus (L.)

Family: Fabaceae

Origin: Papua New Guinea.

Edible parts: Young tender pods, young leaves and shoots, flower, dry seeds and immature tuberous roots. Importance

Propagation: by seeds

Yield: 10-15t/ha.

Globe artichoke

Cynara scolymus

Family: Compositae

Origin: Europe

Edible part: Immature flower buds bracts

Uses: leaves are useful in diuretic and reduction of some toxic substances

Propagation: Seeds & Suckers

Harvesting: 7-8 month after sowing

Yield: 10-12 t/ha.

YAM BEANS

Pachyrrhizus erosus

Family: Leguminoceae native North America distribution China, Thailand, Singapore, India- it is grown in Bihar, West Bengal and Orissa.

Uses: Mature seeds high content of alkaloids young tender pods are used as vegetable.

Propagation: By seed

Harvesting: December-January

Yield: 18-20 t/ha

LARGER YAM

Dioscorea alata

Family: Dioscoreaceae

Origin: South East Asia

Distribution: India, Malaysia, Indonesia and some other parts of Asia

Important Species: *D. esculenta, D. rotundata, D. bulbifera.*

Propogation: small tuber,

Harvesting: 8-9 month after planting

Uses: Unprocessed food forms: Boiled yam, Fried yam and Baked yam.

Processed food forms and value added products: Yam flour, Yam flakes, and Yam chips. feed for livestock.

SNAP MELON

Cucumis melo var. *momordica*

Family: Cucurbitaceae

Growing region: Rajasthan, Gujrat, Punjab, U.P and M.P.

Edible part: Immature fruits and ripe fruits.

Uses: Immature fruits used as salads, vegetables and for rayata making. Ripe fruits used as desert and can be dried & stored for future use.

Propagation: by seeds

Yield: summer season crop: 17-20 t/ha, rainy season crop: 20-25 t/ha

TREE TOMATO

Cyphamandra betacea

Family: Solanaceae

Origin: Peruvian Andes.

Edible part: Ripe fruits.

Uses: Ripe fruits are eaten as raw, salad, and as cooked vegetable. Fruits have pleasant flavor and used to prepare chutney, jam, sauce, and jelly.

Propagation: by seed or through cutting

Yield: 10-15 kg/plant.

REFERENCES

1. ADA (American Diabetes Association). Clinical practice recommendations Diabetes Care. 2016; 20(30):S16.
2. Attanayake AP, Arawwawala LDAM, Jayatilaka KAPW. Chemical standardization of leaf extract of Coccinia grandis (L.) Voigt (Cucurbitaceae) of Sri Lankan origin, Journal of Pharmacognosy and Phytochemistry. 2016; 5(5):119-123.
3. Attanayake AP, Jayatilaka KAPW, Pathirana C, Mudduwa LKB. Acute hypoglycemic and antihyperglycemic effects of ten Sri Lankan medicinal plant extracts in healthy and streptozotocin induced diabetic rats. International Journal of Diabetes in Developing Countries. 2014. doi.10.1007/s13410-014-0217-8.
4. Attanayake AP, Jayatilaka KAPW, Pathirana C, Mudduwa LKB: Study of antihyperglycaemic activity of some Sri Lankan medicinal plants in alloxan induced diabetic rats. Ancient Science of life. 2013; 32(4):193-8.
5. Aung L.H, Harris C.M, Jenner J.F (2004). Chemical growth regulators on postharvest sprout development of *Sechium edule* Swartz. Int J Exper Bot 53: 155-164.
6. Bhattacharya M. A historical exploration of Indian diets and a possible link to insulin resistance syndrome. Appetite. 2015; 95:421-454.
7. Cho SS, Qi L, Fahey GCr, Klurfeld DM. Consumption of cereal fiber, mixtures of whole grains and bran, and whole grains and risk reduction in type 2 diabetes, obesity, and cardiovascular disease. American Journal of Clinical Nutrition. 2013. http://dx.doi: 10.3945/ajcn.113.067629.
8. Cubero, J.I. 1974. On the evolution of *Vicia faba* L. Theor. Appl. Genet. 45: 47-51.
9. Deivanai S. and Bhore Subhash J. Breadfruit (Artocarpus altilis Fosb.)- An Underultilized and Neglected Fruit Plant Species. *Middle-east Journal of Scientific Research.*
10. Ediriweera ERHSS, Ratnasooriya WD. A review on herbs used in treatment of diabetes mellitus by Sri Lankan Ayurvedic and traditional physicians. Ayu. 2009; 30:373- 391.

11. Ekor E: The growing use of herbal medicines: Issues relating to adverse reactions and challenges in monitoring safety. Frontiers in Pharmacology. 2013. http://doi: 10.3389/fphar.2013.00177.

12. Food and Agriculture Organization of the United States – FAO, 2009. FAOSTAT Agriculture data. FAO. Available from: . Access in: 16 mar. 2009.

13. Forbes JM, Cooper ME. Mechanisms of diabetic complications. Physiological reviews. 2011; 93(1):137- 188.

14. Gopalan, C., Rama Sastri, B.V. and Balasubramanian, S.C. Nutritive value of Indian foods. National Institute of Nutrition, Indian Council of Medical Research, Hyderabad, 2012.

15. Jayaweera DMA. Medicinal Plants (indigenous and exotic) used in Ceylon. 2nd ed. Sri Lanka, Colombo, 1982.

16. Jordan SA, Cunningham DG, Marles RJ. Assessment of herbal medicinal products: Challenges and opportunities to increase the knowledge base for safety assessment. Toxicology and Applied Pharmacology. 2010; 243(2):198-216.

17. Lebot, V., 2009. Tropical root and tuber crops: cassava, sweet potato, yams and aroids. Crop production science in horticulture series: 17. Wallingford: CABI. 432 p.

18. Ley SH, Hamdy O, Mohan V, Hu FB. Prevention and management of type 2 diabetes: Dietary components and nutritional strategies. The Lancet. 2014; 383(9933):1999- 2007.

19. Medagama AB, Bandara R, Abeysekera RA, Imbulpitiya B, Pushpakumari T. Use of complementary and alternative medicines (CAMs) among type 2 diabetes patients in Sri Lanka: A cross sectional survey. BMC Complementary and Alternative Medicine. 2014. http://doi.10.1186/1472-6882-14-374.

20. Munasinghe MAAK, Abeysena C, Yaddehige IS, Vidanapathirana T, Piyumal KPB. Blood sugar lowering effect of Coccinia grandis (L.) J. Voigt: Path for a new drug for diabetes mellitus. Experimental Diabetes Research. 2011. http://dx.doi.org/10.1155/2011/978762.

21. Ndong M, Uehara M, Katsumata S, Suzuki K. Effects of oral administration of Moringaoleifera Lam on glucose tolerance in Goto-Kakizaki and Wistar Rats. Journal of Clinical Biochemistry and Nutrition. 2007; 40(3):229- 233.

22. Odebunmi, EO., OLUWANIYI, OO., SANDA, AM. and

KOLADE, BO., 2007. Nutritional compositions of selected tubers and root crops used in Nigerian food preparations. International Journal of Chemistry, vol. 17, no. 1, p. 37-43.

23. Partap, T., Joshi, B.D. & Galwey, N., 1998. Chenopods, Chenopodium spp. Promoting the conservation and use of underutilized and neglected crops 22. International Plant Genetic Resources Institute (IPGRI), Rome, Italy. 67 pp.

24. Pedralli, G., 2002. Distribuição geográfica e taxonomia das famílias Araceae e Dioscoreaceae no Brasil. In CARMO, CAS. (Ed.). Inhame e taro: sistemas de produção familiar. Vitória: Incaper. p. 15-26.

25. Petrus, A.J.A. Sauropus androgynus (L.) Merrill- A potentially nutritive functional leafy vegetable. *Asian J. Chem.*, 2013, 25, 9425–9433.

26. Post RE, Mainous AG, King DE, Simpson KN. Dietary fiber for the treatment of type 2 diabetes mellitus: A meta-analysis. Journal of American Board of Family Medicine. 2012; 25(1): 16-23.

27. Raina, S.N. and Ogihara, Y. 1995. Ribosomal DNA repeat unit polymorphism in 49 Vicia species. *Theor. Appl. Genet.* 90: 477-486.

28. Ram, D.; Kumar, S.; Verma, A. and Rai, M. (2004). Variability analysis of underutilized nutritive vegetable kartoli. Cucurbit Genet. Coop. Rpt. 27: 66-68.

29. Ramaswamy, N., muthukrishnan, C. R. and Shanmucavelu, K. G. 1980. Varietal preference of mishrikand. National Seminar on Tuber Crop Production Technology, (India, 1980), pp. 119-200. Coimbatore, India: Tamil Nadu Agricultural University.

30. Ramsay, G.R. and Pickersgill, B. 1986. Interspecific hybridisation between *Vicia faba* and other species of *Vicia:* approaches delaying embryo abortion. *Biol. Zentralbl.* 105: 171-179.

31. Rubatzky, V.E. & Yamaguchi, M., 1997. World vegetables: principles, production and nutritive values. 2nd Edition. Chapman & Hall, New York, United States. 843 pp.

32. Saade R.L (1996). Chayote. Sechium edule (Jacq.) Sw.: Promoting the conservation and use of underutilized and neglected crops. International Plant Genetic Resources Institute,

Rome, Italy.

33. Sheela, K., Nath, K.G., Vijayalakshmi, D., Yankanchi, G.M. and Patil, R.B. Proximate composition of underutilized green leafy vegetables in Southern Karnataka. *J. Hum. Ecol.*, 2004, 15, 227–229.

34. Singh B.K, Pathak K.A, Ngachan S.V (2012). Exploring underutilized chow-chow in Mizoram. Indian Hort 57 (5): 3-5.

35. Singh B.K, Pathak K.A, Ramakrishna Y (2013). Underutilized vegetable crops and spices of Mizoram: needs exploration and utilization. In: Prakash N, Roy S.S, Sharma P.K, Ngachan S.V (Eds) Developing the potential of underutilized horticultural crops of hill regions. Today & Tomorrow's Printers and Publishers, New Delhi, pp 217-232.

36. Tindall, H. D. 1968. Bean yam; potato bean. Commercial vegetable growing, pp. 117-118. London: Oxford University Press, 300 pp.

37. Ukom, A.N. and Okerue, C.F.L. 2005. Determination of the Nutricnts, Anti-Nutricnts and Functional Properties of Processed Cocoyam (Xanthosoma sagittifolium) Cultivars Grown in Southeast, Nigeria. *Sustainable Food Production.* Vol. 1, pp 11-21.

38. Verma VK, Jha A.K and Singh B.K (2014). Nutritional properties of different fruit parts of popular chow-chow genotype grown in NEH Region of India. Vegetable Newsletter 1 (1): 8.

39. Zimmet PZ, Magliano DJ, Herman WH, Shaw JE: Diabetes: A 21st century challenge. Lancet Diabetes Endocrinology. 2014; 2(1):56-64.

40. Zohary, D. and Hopf, M. 1973. Domestication of pulses in the old world. *Science* 182: 887-894.

41. Nieuwhof, M. 1969. *Cole Crop.* World Crop Book, Leonard Hill. pp 96- 101.

42. Robbins, W. W. 1931. *The Botany of Crop Plant.* P. Blakiston's Son. Kholrabi. pp 321- 322.

43. Yamaguchi, M. 1963. *World vegetables. pp 223- 224.*

44. Thamburaj, S. and Narendra, Singh, 2005. *Kholrab in Textbook of Vegetables, tuber and Spices.* pp 141- 146.

Related Books

ISBN: 9788172339029

ISBN: 9788172336189

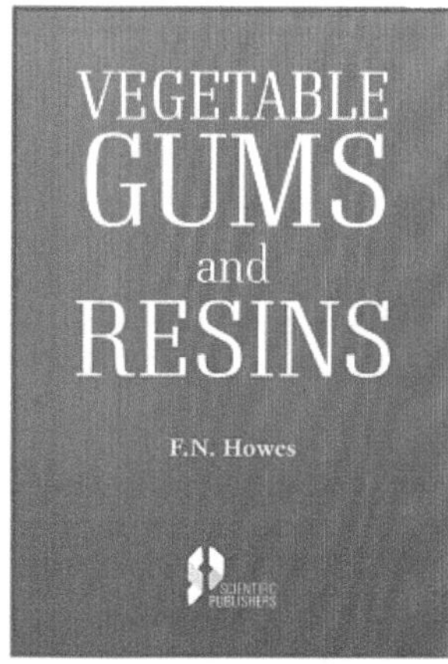

ISBN: 9788172332723

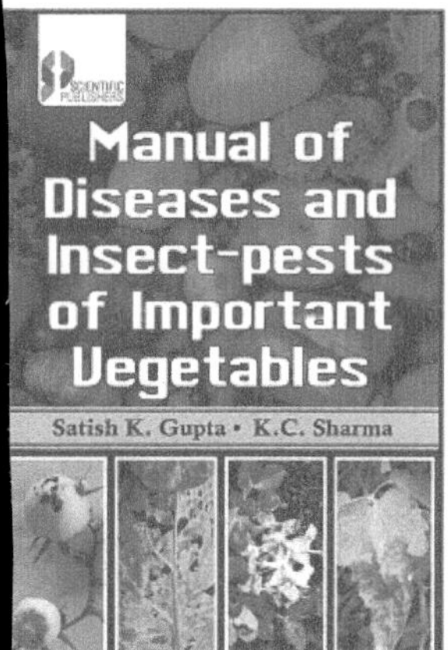

ISBN: 9789389412291

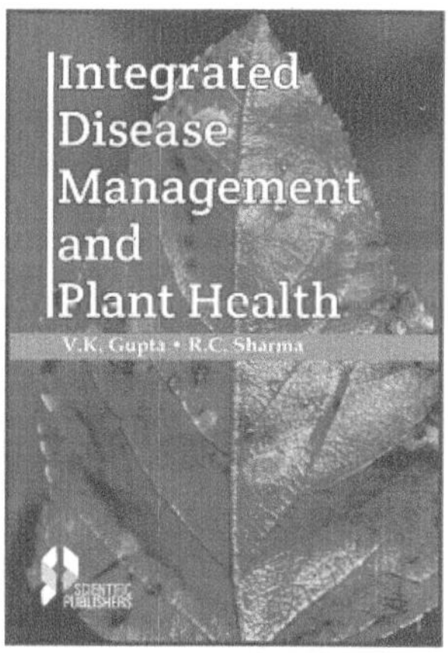

ISBN: 9789386102553

ISBN: 9789389412277

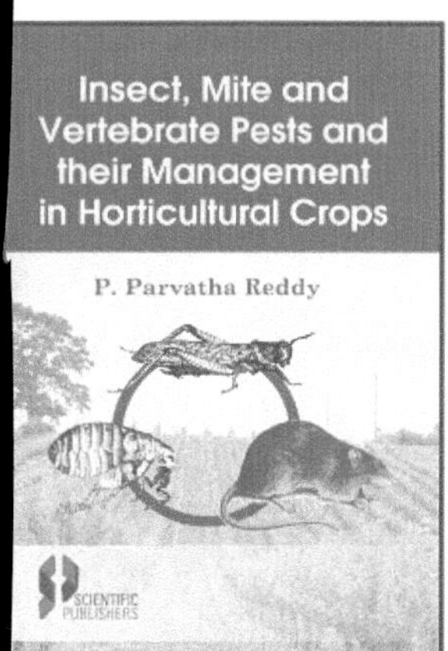

ISBN: 9788172336288

ISBN: 9789387741249

ISBN: 9789386102812

For Full Content & Sample Page Visit Website

Related Books

ISBN: 9789388043649

ISBN: 9789386652232

ISBN: 9789387307087

ISBN: 9789388043298

ISBN: 9789386652348

ISBN: 9789386102669

ISBN: 9789386652386

ISBN: 9789386652478

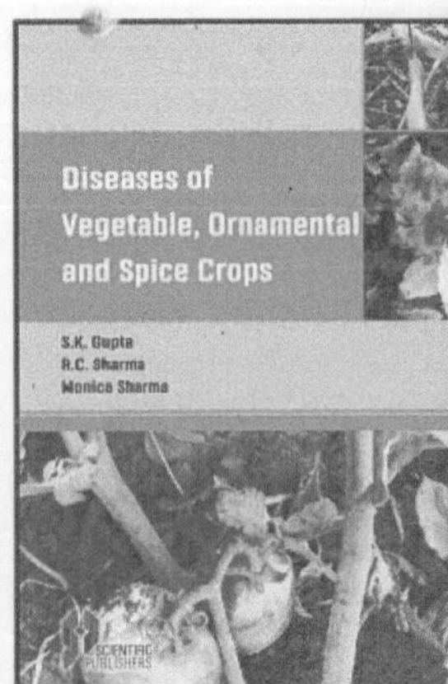

ISBN: 978938730716

For Full Content & Sample Page Visit Website